Instructor's Manual with Te:

Manese Rabeony

Rutgers University

to accompany

Foundations of College Chemistry

11th Edition
Alternate 11th Edition

Morris Hein
Mount San Antonio College

Susan Arena
University of Illinois
Urbana–Champaign

WILEY

JOHN WILEY & SONS, INC.

To order books or for customer service call 1-800-CALL-WILEY (225-5945).

Contents

CHAPTER 1 TEST

True — False

_____ 1. A key feature of the scientific method is to plan and do additional experiments to test a hypothesis.

_____ 2. Chemistry is the science that deals with the composition of substances and the transformation they undergo.

_____ 3. An explanation of many observations that have been proven by many tests is called a hypothesis.

_____ 4. One who assesses chemical risk is not really concerned about the probability of exposure to a particular chemical, only the severity of the potential exposure.

_____ 5. Chemistry has applications in the fields of agriculture, pharmaceuticals, and environmental science, among others.

Multiple Choice (choose the best answer)

_____ 1. Chemists often repeat the experiments of other scientists to verify or check results. Which process of science does this illustrate?

a) classification   b) communicating   c) hypothesizing   d) interpreting

_____ 2. The science of chemistry does or can involve

a) observations of nature   b) product development   c) risk assessment   d) a, b, and c

_____ 3. If scientific experiment produces consistent results over a long period of time, it would verify a

a) theory   b) hypothesis   c) law   d) rule

_____ 4. Before any idea or hypothesis can be presented, a scientist must gather significant

a) ideas   b) scientists   c) data   d) equipment

_____ 5. The language of chemistry involves

a) symbols   b) laws   c) theories   d) a, b, and c

_____ **6.** Which of the following is a scientific observation?

    a) Freezing and boiling are called physical changes.
    b) If a substance freezes at 0°C, it must be water.
    c) When a substance freezes, its molecules lose potential energy.
    d) Water freezes at 0°C.

Reasoning and Expression (use a separate sheet of paper)

1. Identify three ways in which Buehler's development of memory metal illustrates the scientific method at work.

2. In what way is it important that *testing* be the link between a hypothesis and a theory? Between a theory and a law?

3. Suppose someone has just brought you a brand new material. Give five questions that you, the chemist, would likely ask about it in an attempt to discover what it is.

4. Why are both severity and probability of risk of exposure to a certain chemical important in assessing overall risk?

5. Restate the four steps in the scientific method.

6. Name four different ways in which you could classify or organize twenty different plants.

CHAPTER 2 TEST

True — False

_____ **1.** As a metric prefix, *kilo*, means 1,000 or 10^3.

_____ **2.** The joule, like the calorie, measures temperature.

_____ **3.** The number 14.077675 rounded off to five digits is 14.078.

_____ **4.** The answer to 25.5 x 0.1678 should contain three significant digits.

_____ **5.** As a material is heated, its volume increases. If its mass stays the same during the heating process, its density will decrease.

_____ **6.** Two cubes of the same size but different masses will have different densities.

_____ **7.** A material whose specific gravity is 1.4 will float in water.

_____ **8.** The number 0.002040 written in scientific notation is 2.04×10^{-2}.

_____ **9.** At –20°C, the Fahrenheit and Celsius temperatures are equal.

_____ **10.** One millimeter is exactly equal to 1 cm^3.

_____ **11.** The SI system of measuring is convenient to scientists because of its dependence on factors of 10.

Multiple Choice (choose the best answer)

_____ **1.** 1.00 cm is equal to how many meters?

　　a) 2.54　b) 100.　c) 10.0　d) 0.0100

_____ **2.** Express 0.00382 in scientific notation.

　　a) 3.82×10^3　b) 3.82×10^{-2}　c) 3.82×10^{-3}　d) 3.8×10^{-3}

_____ **3.** 267°F is equivalent to

　　a) 404 K　b) 116°C　c) 540 K　d) 389 K

_____ **4.** The mass of a block is 9.43 g and its density is 2.35 g/mL. The volume of the block is

a) 4.01 mL   b) 22.2 mL   c) 0.249 mL   d) 2.49 mL

_____ **5.** The density of copper is 8.92 g/mL. The mass of a piece of copper that has a volume of 9.5 mL, is

a) 2.58 g   b) 85 g   c) 0.94 g   d) 107 g

_____ **6.** An empty graduated cylinder has a mass of 54.772 g. When filled with 50.0 mL of an unknown liquid it has a mass of 101.074 g. The density of the liquid is

a) 0.926 g/mL   b) 2.02 g/mL   c) 1.00 g/mL   d) 1.845 g/mL

_____ **7.** All of the following units can be used for density except

a) g/cm^3 b) kg/m^3 c) g/L   d) kg/m^2

_____ **8.** 37.4 cm x 2.2 cm equals

a) $82.28 \, cm^2$ b) $82.3 \, cm^2$ c) $82 \, cm^2$ d) $82.2 \, cm^2$

_____ **9.** The number 0.0048 contains how many significant digits?

a) 2   b) 3   c) 4   d) 5

_____ **10.** 4.50 feet is how many centimeters?

a) 11.4   b) 21.3   c) 454   d) 137

_____ **11.** All of the following except which is an accepted SI unit?

a) kilogram   b) meter   c) pound   d) Kelvin

Reasoning and Expression (use a separate sheet of paper)

1. If the density of gasoline is 0.68 g/cm^3, what is the mass of 10.0 gallons of it?

2. Of these three, which is the lowest temperature: 84°C, 200°F, or 350 K?

3. What is the density of a rectangular block of wood if it measures 4.0 cm x 12 cm x 20.0 cm and has a mass of 720 grams? Will it sink or float in water? Explain briefly.

4. Give the appropriate SI unit to measure each of the following:

a) thickness of a penny
b) amount of liquid in a soda can
c) height of a redwood tree
d) heat energy contained in a boiler of water
e) surface area of a large ballroom

5. Explain how an ocean of water and a cup of that same ocean water can have the same temperature but contain different amounts of heat.

6. Here are some data gathered by three experiments:

    30 buttons weigh 20.8 g
    120 buttons 82.0 g
    50 buttons weigh 35.8 g

    a) Plot the data.
    b) According to your graph, predict the number of buttons that would equal 75.0 g.
    c) What is the average density of the buttons in gram/button?

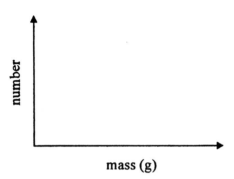

CHAPTER 3 TEST

True — False

_____ **1.** Matter that has identical properties throughout is homogenous.

_____ **2.** A gas is the least compact of the three states of matter.

_____ **3.** Most elements occur in nature as metallic solids.

_____ **4.** A molecule is a small, uncharged individual unit of a compound formed by the union of two or more atoms.

_____ **5.** The symbol for silver is S.

_____ **6.** An ion is an electrically charged atom or group of atoms.

_____ **7.** A liquid has a definite shape and a definite volume.

_____ **8.** To express a molecule of phosphorus which is composed of four atoms, one could accurately use either the symbol P_4 or $4P$.

_____ **9.** All gases exist diatomically.

_____ **10.** The main characteristic of a mixture is definite composition.

_____ **11.** Different compounds may contain the very same elements in different proportions.

Multiple Choice **(choose the best answer)**

_____ **1.** Which of the following is a compound?

   a) water   b) potassium   c) wood   d) lead

_____ **2.** How many atoms are represented in the formula Na_2SO_4?

   a) 3   b) 5   c) 7   d) 8

_____ **3.** Which of the following is a characteristic of all nonmetals?

   a) always gases   b) poor conductors of electricity
   c) shiny in appearance   d) combine only with metals

7

_____ **4.** Which of the following is an amorphous solid?

a) quartz   b) glass   c) silver   d) sodium

_____ **5.** The number of nonmetallic atoms in $Al_2(SO_3)_3$ is

a) 5   b) 7   c) 12   d) 14

_____ **6.** If element X is known to be a metal, one would expect it to be

a) ductile   b) extremely strong   c) easily shattered   d) dull in appearance

_____ **7.** Chromium, flourine, and magnesium have the symbols

a) Ch, F, Ma   b) Cr, F, Mg   c) Cr, Fl, Mg   d) Cr, F, Ma

_____ **8.** Coffee is an example of

a) an element   b) a compound   c) a homogenous mixture   d) a heterogeneous mixture

_____ **9.** All of the following characterize solutions, except

a) They are homogenous mixtures   b) They are heterogenous mixtures
c) They contain two or more substances   d) They have variable composition

_____ **10.** A state of matter which lacks a definite volume is

a) solid   b) liquid   c) gas   d) amorphous substance

_____ **11.** Which of the following is an example of a chemical change?

a) A mixture of baking soda and vinegar fizzes vigorously.
b) A heated candle turns to a liquid.
c) Ice changes into water.
d) A sheet of copper is easy to bend and fold.

_____ **12.** Which would you classify as a semiconductor?

a) diamond   b) silicon   c) lead   d) $MgBr_2$

Reasoning and Expression (**use a separate sheet of paper**)

1. List the following compounds in order of increasing number of oxygen atoms represented:
     12 $CaCO_3$ 8 $Ca(OH)_2$ 6 $Ca_3(PO_4)_2$ $Ca(NO_3)_2$

2. Differentiate briefly between

a) an atom and an ion   b) an anion and a cation
c) a solid and a gas   d) an atom and a molecule

3. You are given two white powders. Identify three things you would do to begin to chemically identify them.

4. From the diagrams given below, identify which shows:

a) only uncombined atoms   b) a sample of a diatomic gaseous element   c) a mixture

 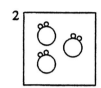 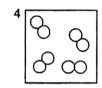

5. In these compounds, identify the ratio of hydrogen atoms to all atoms

a) CH_4 b) $C_{25}H_{52}$ c) $C_6H_{12}O_6$ d) CH_3COOH

6. Identify an example of an element which fits the following descriptions:
a) a monatomic gas
b) a diatomic liquid
c) a metallic liquid
d) a metal which can be found in nature in the free state
e) the metallic element in common table salt

CHAPTER 4 TEST

True — False

_____ 1.   In a physical change, substances are formed that are entirely different, having different properties and composition from the original material.

_____ 2.   The Law of Conservation of Mass states that no detectable change is observed in the total mass of the substances involved in a chemical change.

_____ 3.   4.184 calories is the equivalent of 1.0 joule of energy.

_____ 4.   The specific heat of water is high compared to most other substances.

_____ 5.   The ending materials in a chemical reaction are called the products.

_____ 6.   A high specific heat indicates that a material requires very little heat to warm up.

_____ 7.   Melting and boiling are examples of physical changes.

_____ 8.   Iron (specific heat = 0.473 J/g°C) will absorb more heat energy per gram than gold (specific heat = 0.0131 J/g°C).

_____ 9.   Energy is the capacity of matter to do work.

_____ 10. Temperature and heat energy are both measured in Kelvins.

_____ 11. Two substances are mixed in a container and the temperature of the system drops dramatically as they combine. In this reaction, energy is being absorbed by the substances involved.

Multiple Choice (choose the best answer)

_____ 1.   Which of the following is not a physical property?

   a) boiling point   b) bleaching action   c) physical state   d) color

_____ 2.   Which of the following is a physical change?

   a) a piece of sulfur is burned   b) a firecracker explodes
   c) a rubber band is stretched   d) a nail rusts

_____ **3.** Which of the following is a chemical change?

  a) water evaporates   b) ice melts   c) rocks are ground to sand   d) a penny tarnishes

_____ **4.** Changing hydrogen and oxygen into water is a

  a) physical change   b) chemical change
  c) conservation reaction   d) None of the above.

_____ **5.** Barium iodide, BaI_2, contains 35.1% barium by mass. An 8.50 g sample of barium iodide contains what mass of iodine?

  a) 5.52 g   b) 2.98 g   c) 3.51 g   d) 6.49 g

_____ **6.** To heat 30. g of water from 20.°C to 50.°C will require

  a) 30 calories   b) 50 calories   c) 3.8×10^3 J   d) 6.3×10^3 J

_____ **7.** The specific heat of aluminum is 0.900 J/g°C. How many joules of energy are required to raise the temperature of 20.0 g of Al from 10.0°C to 15.0°C?

  a) 79 J   b) 90.0 J   c) 100. J   d) 112 J

_____ **8.** A 100. g iron ball (specific heat = 0.473 J/g°C) is heated to 125°C and placed in a calorimeter holding 200 g of water at 25.0°C. What will be the highest temperature reached by the water?

  a) 43.7°C   b) 30.4°C   c) 65.3°C   d) 35.4°C

_____ **9.** Which of these is *not* a chemical change?

  a) heating of copper in air   b) combustion of gasoline
  c) cooling of red-hot iron   d) digestion of food

_____ **10.** Which of these is *not* involved in the calculation of heat absorbed as a substance is warmed?

  a) mass of the sample   b) change in temperature
  c) density of the sample   d) specific heat of the material

Reasoning and Expression (use a separate sheet of paper)

1. A 12 g strip of copper metal is held in a container of iodine vapor. Gradually, the strip is covered with the white compound, CuI. The strip now has a mass of 36 g. Describe how to determine the percent of iodine in the compound.

2. A 25.0 g sample of lead (specific heat = 0.128 J/g°C) was warmed from 35°C to 200.°C. Now that same amount of energy is applied to a 60.0 sample of Cu (specific heat = 0.385 J/g°C) at 35°C. Determine the final temperature of the copper.

3. Translate the following words into chemical equations:

   a) calcium and water mix to form calcium hydroxide, $Ca(OH)_2$, and hydrogen, H_2
   b) iron and sulfur combine to form iron (II) sulfide, FeS
   c) carbon burns in oxygen gas to form carbon dioxide

4. Imagine the following scene: A golfer takes his golf club and swings it back, preparing to hit the ball. The club stops momentarily at the top of the swing. Then, the golfer vigorously swings it down, striking the ball and taking the club back up on his follow-through. Describe the changes in potential and kinetic energy that occur in the event. At what point is the club's potential energy at its maximum? At what point is the kinetic energy at its maximum?

5. Compare a substance with a high specific heat to a substance with a low specific heat with respect to ability to absorb or release heat.

6. If this is a model representing a solid, draw a similar diagram representing a liquid and a gas.

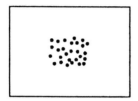

7. Three different substances are placed in three separate, identical, open dishes and weighed. They are each heated for a period of time. Following their heating and cooling, one of the dishes is found to have lost mass, one is found to have gained mass, and one is found to have exactly the same mass it had before heating. How do you explain these results, knowing that mass is conserved in chemical reactions?

CHAPTER 5 TEST

True — False

_____ **1.** Dalton's atomic theory states that all atoms are composed of protons, neutrons, and electrons.

_____ **2.** The electron was discovered by J.J. Thomson.

_____ **3.** The proton and the neutron have approximately equal charge and mass.

_____ **4.** The atoms of two different elements must have different mass numbers.

_____ **5.** The lightest element is helium.

_____ **6.** The nucleus of an atom contains protons, neutrons, and electrons.

_____ **7.** Emission of an alpha particle alters the mass and charge of the nucleus.

_____ **8.** The listed atomic mass of an element represents the relative average mass of all the naturally occurring isotopes of that element.

_____ **9.** An atom of silver-108 contains 47 protons, 47 electrons, and 108 neutrons.

_____ **10.** The reason that the atomic mass of Mg is 24.305 rather than almost exactly 24 is that protons and neutrons do not have exactly the same mass.

_____ **11.** An atom of Mg-27 and an ion of P-31 (with a +3 charge) contain equal numbers of electrons, though the ion would have a greater mass.

Multiple Choice **(choose the best answer)**

_____ **1.** The concept of most of an atom's mass being concentrated in a small nucleus surrounded by the electrons was the contribution of

a) Dalton   b) Thomson   c) Rutherford   d) Chadwick

_____ **2.** An atom of atomic number 53 and mass number 127 has how many neutrons?

a) 53   b) 127   c) 74   d) 180

_____ **3.** How many electrons are in an atom of argon-40?

a) 20   b) 40   c) 22   d) No correct answer given.

_____ **4.** The protons and neutrons in an atom always add to give the atom's

     a) atomic number   b) ionic charge   c) number of electrons   d) mass number

_____ **5.** Each atom of a specific element has the same

     a) number of protons   b) atomic mass   c) number of neutrons   d) nuclear mass

_____ **6.** Isotopes differ from one another in all but which of the following ways?

     a) number of protons   b) number of neutrons   c) nuclear mass   d) atomic mass

_____ **7.** Substance X has 13 protons, 14 neutrons, and 10 electrons. Determine its identity:

     a) ^{27}Mg b) ^{27}Ne c) $^{27}Al^{3+}$ d) ^{27}Al

_____ **8.** The mass of a chlorine atom is 5.9×10^{-23}g. How many atoms are in a 42.0 sample of chlorine?

     a) 2.48×10^{-21} b) 1.40×10^{-24} c) 7.12×10^{23} d) No correct answer given.

_____ **9.** The number of protons in an atom of zinc-65 is

     a) 65   b) 30   c) 35   d) 95

_____ **10.** Consider particle A with a molar mass of 62, atomic number 32, and ionic charge of +2. Which of these is correct?

     a) A has an equal number of p and n   b) A has an equal number of p and e^{-}
     c) A has an equal number of n and e^{-} d) A has an equal number of p, n, and e^{-}

_____ **11.** Which part of John Dalton's theory did Thomson's findings refute?

     a) Atoms contain positive and negative particles   b) Atoms are indivisible
     c) Different atoms have different masses   d) Atoms are very small

Reasoning and Expression (use a separate sheet of paper)

    1. Briefly describe the differences among models of the atom devised by Dalton, Thomson and Rutherford as far as location of the atom's positive charge is concerned.

    2. Specifically cite experimental evidence that proves

       a) the nucleus of an atom is positively charged
       b) the nucleus of an atom is highly dense
       c) the atom is mostly empty space

3.   A mystery element Q occurs as three isotopes. Analysis of a sample of Q showed:

Isotope	Mass	Percent
1	80.0 amu	60.0
2	84.0 amu	30.0
3	82.0 amu	10.0

Determine the mass number which should be listed on the periodic table.

4.   Explain why most elements on the periodic table have mass numbers that are very close to being whole numbers.

5.   What is the relationship between an atom which contains 10 protons, 10 electrons, and 11 neutrons and one which contains 11 protons, 11 electrons, and 10 neutrons?

6.   How would the resulting observations in Rutherford's gold-foil experiment have been different if alpha particles were negatively rather than positively charged?

CHAPTER 6 TEST

True — False

_____ **1.** Binary compounds are denoted by the suffix *ide*.

_____ **2.** The name of $ZnBr_2$ is zinc bromide.

_____ **3.** An ion is a positive or negative electrically charged atom or group of atoms.

_____ **4.** The formula for barium hydroxide is BaOH.

_____ **5.** The formula for copper (II) oxide is Cu_2O.

_____ **6.** The prefixes *tri*, *tetra*, and *penta* represent 2, 3, and 4, respectively.

_____ **7.** The common name for sulfur is brimstone.

_____ **8.** The formula for muratic acid is HNO_3.

_____ **9.** The compound formed from Ga^{3+} and oxygen would be Ga_3O_2.

_____ **10.** The formula for ammonium hydroxide is NH_4OH.

_____ **11.** The formula for dinitrogen pentoxide is N_2O_5.

_____ **12.** The formula for ferrous sulfide is FeS_2.

_____ **13.** The formula for sodium carbonate is $NaCO_3$.

_____ **14.** The formula for ferric iodide is FeI_2.

_____ **15.** The formula for Hg(I) iodide is HgI.

_____ **16.** The formula for tin (IV) nitrate is $Sn(NO_3)_4$.

_____ **17.** The formula for barium iodide is Ba_2O_2.

_____ **18.** The name of $ZnBr_2$ is zinc bromide.

_____ **19.** The name of $CuCl_2$ is cupric chloride.

_____ **20.** The name of P_2O_5 is diphosphorous tetroxide.

_____ **21.** In order for an ion to have "-ate" as its suffix, that ion must contain the element hydrogen.

_____ **22.** Molecules of the compounds calcium oxide, magnesium oxide, and aluminum oxide all contain the same number of atoms.

Names and Formulas

Write names for:

1. H_2SO_4 _____

2. HNO_3 _____

3. $HC_2H_3O_2$ _____

4. $Cu(OH)_2$ _____

5. $Fe_2(SO_4)_3$ _____

6. $MgCO_3$ _____

7. SnS_2 _____

Write formulas for:

1. hydrochloric acid _____

2. phosphoric acid _____

3. potassium hydroxide _____

4. sodium sulfate _____

5. aluminum chloride _____

6. copper (I) oxide _____

7. ammonium bromide _____

Reasoning and Expression (use a separate sheet of paper)

1. Translate the following sentences into chemical equations:

a) barium chloride and sulfuric acid combine to form barium sulfate and hydrochloric acid
b) sodium carbonate and lead (II) nitrate yield sodium nitrate and lead (II) carbonate
c) cupric nitrate and sodium hydroxide yield cupric hydroxide and sodium nitrate

2. Write the formulas that you would expect for compounds formed between:

a) Q^{5+} and Z^{2-} b) A^{1+} and B^{4-} c) ABC^{2+} and XYZ^{3-}

3. a) Where on the periodic table do you find elements that tend to form positively charged ions?
b) Where are the elements located that tend to form negatively charged ions?

4. What is true of the ions formed by beryllium, magnesium, calcium, strontium, and barium? Where are they located on the periodic table? What is true of the ions formed by elements found one column to their left on the table?

5. Identify at least four different elements that can form more than one kind of cation.

6. Nitrogen can combine with oxygen in a number of ways; some of the compounds that can result include NO, NO_2, N_2O, and N_2O_4. In all cases, each of the oxygen atoms carries a –2 charge. What would be the charge on the nitrogen in each compound?

CHAPTER 7 TEST

True — False

_____ **1.** A mole of Ag(107.9 g) contains the same number of atoms as a mole of Na(22.99 g).

_____ **2.** One mole of glucose, $C_6H_{12}O_6$, contains 24 mol of atoms.

_____ **3.** One gram of sulfur contains 6.022 x 10^{23} atoms.

_____ **4** The mass of a mole of NaCl is less than the mass of a mole of KCl.

_____ **5.** The percent composition of a compound is the mass percent of each element in the compound.

_____ **6.** The empirical formula of a compound gives the smallest ratio of the atoms that are present in the compound.

_____ **7.** A mole of magnesium and a mole of magnesium oxide, MgO, contain the same number of magnesium atoms.

_____ **8.** The number of sulfur atoms is the same in 1 mol of Na_2SO_4 as in 1 mol of K_2SO_4.

_____ **9.** There are 14 mol of chlorine atoms in 3.5 mol of CCl_4.

_____ **10.** A compound that has a carbon to hydrogen ratio of 1:2 can have a molar mass of 48.0.

_____ **11.** If an oxide of iron contains 30% oxygen, the formula of the oxide must be Fe_2O_3.

Multiple Choice **(choose the best answer)**

1. Avogadro's number of magnesium atoms

a) Have a mass of 1.0 g   b) Have the same mass as Avogadro's number of sulfur atoms
c) Have a mass of 12.0g   d) Is 1 mol of magnesium atoms

2. The number of moles in 112g of acetylsalicylic acid (aspirin), $C_9H_8O_4$, is

a) 1.61   b) 0.622   c) 112   d) 0.161

3. How many grams of Au2S can be obtained from 1.17 mol of Au?

a) 182 g   b) 249 g   c) 364 g   d) 499 g

_____ **4.** A 16 g sample of O_2

a) Is 1 mol of O_2 b) Contains 6.022 x 10^{23} molecules of O_2
c) Is 0.50 molecule of O_2 d) Is 0.50 molar mass of O_2

_____ **5.** 2.00 mol of CO_2

a) Have a mass of 56.0 g   b) Contain 1.20 x 10^{24} molecules
c) Have a mass of 44.0 g   d) Contain 6.00 molar masses of CO_2

_____ **6.** In Ag_2CO_3, the percent by mass of

a) Carbon is 43.5%   b) Silver is 64.2%   c) Oxygen is 17.4%   d) Oxygen is 21.9%

_____ **7.** A compound contains 54.3% C, 5.6% H, and 40.1% Cl. The empirical formula is

a) CH_3Cl b) C_2H_5Cl c) $C_2H_4Cl_2$ d) C_4H_5Cl

_____ **8.** A compound contains 40.0% C, 6.7% H, and 53.3% O. The molar mass is 60.0 g/mol. The molecular formula is

a) $C_2H_3O_2$ b) C_3H_8O c) CH_2O d) $C_2H_4O_2$

_____ **9.** How many chlorine atoms are in 4.0 mol of PCl_3

a) 3   b) 7.2 x 10^{24} c) 12.0   d) 2.4 x 10^{24}

_____ **10.** The empirical formula of a compound is CH. The molar mass of this compound is 78.0, then the molecular formula is

a) C_2H_2 b) C_5H_{18} c) C_6H_6 d) No correct answer given.

_____ **11.** Which of these samples would have the greatest mass?

a) 58.9 grams cobalt   b) 3.01 x 10^{23} molecules H_2O_2
c) 120 milliliters of salt water (density = 1.3 g/mL)   d) 12.1 moles of ZnO

Reasoning and Expression (use a separate sheet of paper)

1. Which of the following samples weighs the least? Which contains the greatest number of atoms?
a) 0.76 mole tungsten
b) 48.25 grams boron trifluoride
c) 8.8 x 10^{22} atoms aluminium

2. A mystery substance X contains 0.667 mol of nitrogen atoms, 2.688 g of hydrogen, 2.01 x 10^{23} chromium atoms, and half as many oxygen as hydrogen atoms. What is the empirical formula?

3. What is the difference between what density measures and what atomic mass measures? How are they similar?

4. How many carbon atoms contain about the same mass as a titanium atom?

5. Vitamin B-12 has the chemical formula $C_{63}H_{88}CoN_{14}O_{14}P$. Is this formula empirical? Explain.

6. Determine the molar masses of these compounds:
   a) 8.0 g of X contain 4.0×10^{22} molecules
   b) 0.00842 kg of Y contain 203 millimoles of atoms
   c) 27.25 mL of Z (density = 5.82 g/mL) contain 9.4×10^{23} molecules

CHAPTER 8 TEST

True — False

_____ 1. In balancing an equation, we change the formulas of compounds to make the number of atoms on each side of the equation balance.

_____ 2. The equation $N_2 = 3\,H_2 \rightleftarrows 2\,NH_3$ can be interpreted as saying that 1 mol of N_2 reacts with 3 mol of H_2 to form 2 mol of NH_3.

_____ 3. The equation $N_2 = 3\,H_2 \rightleftarrows 2\,NH_3$ can be interpreted as saying that 1 g of N_2 reacts with 3 g of H_2 to form 2 g of NH_3.

_____ 4. The substances on the right side of a chemical equation are called the products.

_____ 5. When a gas is evolved in a chemical reaction, it can be indicated in the equation with the symbol (g) immediately following the formula of the gas.

_____ 6. A balanced chemical equation is one that has the same number of moles on each side of the equation.

_____ 7. The coefficients in front of the formulas in a balanced equation give the relative number of moles of the reactants and products in the reaction.

_____ 8. Water is formed in a neutralization reaction.

_____ 9. The combustion of hydrocarbons is an exothermic reaction.

_____ 10. In a balanced chemical equation, the mass of the products is equal to the mass of the reactants.

_____ 11. Synthesis reactions contain more products than reactants; decomposition reactions contain more reactants than products.

Multiple Choice (choose the best answer)

1. The reaction $BaCl_2 = (NH_4)_2CO_3 \rightarrow BaCO_3 = 2\,NH_4Cl$ is an example of

    a) combination   b) decomposition   c) single displacement   d) double displacement

2. The reaction $2\,Al + 3\,Br_2 \rightarrow 2\,AlBr_3$ is an example of

    a) combination   b) single displacement   c) decomposition   d) double displacement

_____ **3.** When the equation $PbO_2 \xrightarrow{\Delta} PbO + O_2$ is balanced, one of the terms in the balanced equation is

a) PbO_2 b) $3 O_2$ c) $3 PbO$ d) O_2

_____ **4.** When the equation $Cr_2S_3 + HCl \rightarrow CrCl_3 + H_2S$ is balanced, one of the terms in the balanced equation is

a) $3 HCl$ b) $CrCl_3$ c) $3 H_2S$ d) $2 Cr_2S_3$

_____ **5.** When the equation $NH_4OH + H_4SO_4 \rightarrow$ is completed and balanced, one of the terms in the balanced equation is

a) NH_4SO_4 b) $2 H_2O$ c) H_2OH d) $2 (NH_4)_2SO_4$

_____ **6.** When the equation $H_2 + V_2O_5 \rightarrow V +$ is completed and balanced, a term in the balanced equation is

a) $2 V_2O_5$ b) $3 H_2O$ c) $2 V$ d) $8 H_2$

_____ **7.** When the equation $Al(OH)_3 + H_2SO_4 \rightarrow Al_2(SO_4)_3 + H_2O$ is completed and balanced, the sum of the coefficients will be

a) 9   b) 10   c) 11   d) 12

_____ **8.** When the equation $Fe_2(SO_4)_3 + Ba_4(OH)_2 \rightarrow$ is completed and balanced, a term in the balanced equation is

a) $Ba_2(SO_4)_3$ b) $2 Fe(OH)_2$ c) $2 Fe_2(SO_4)_3$ d) $2 Fe(OH)_3$

_____ **9.** For the reaction $2 H_2 + O_2 \rightarrow 2 H_2O + 572.4 kJ$, which of the following is *not* true?

a) The reaction is exothermic.
b) 572.4 kJ of heat are liberated for each mole of water formed.
c) Two moles of hydrogen react with 1 mol of oxygen.
d) 572.4 kJ of heat are liberated for each 2 mol of hydrogen reacted.

_____ **10.** When a nonmetal oxide reacts with water,

a) A base is formed   b) An acid is formed
c) A salt is formed   d) A non metal oxide is formed

_____ **11.** Which of the following is a double-displacement reaction?

a) $(NH_4)_2SO_3 \rightarrow 2NH_3 + H_2O + SO_2$
b) $Br_2 + 2KI \rightarrow 2KBr + I_2$
c) $2Na + Cl_2 \rightarrow 2NaCl$
d) $Al(OH)_3 + 3 HCl \rightarrow AlCl_3 + 3 H_2O$

Reasoning and Expression (use a separate sheet of paper)

1. Suppose that the term 8 $Al_2(SO_4)_3$ appears in a balanced equation. How many atoms of Al, S, O, and total number of atoms is represented?

2. Consider this equation: $X_3Y \xrightarrow{\Delta} 3X + Y$

   a) What kind of reaction does this represent?
   b) Name two changes that occur as this reaction proceeds.
   c) Name two things about this reaction that do not change.

3. Balance these equations:

   a) $Ag + H_2S + O_2 \rightarrow Ag_2S + H_2O$ b) $Al + Fe_3O_4 \rightarrow Al_2O_3 + Fe$

4. Benzene, C_6H_6, is reacted with nitric acid, HNO_3. Two products are formed, one of which is water. The other product is an oily liquid which has a molar mass of 213g/mol. It contains 33.8% C, 1.42% H, 19.7% N, and the remainder oxygen. From this information, write a balanced equation for the reaction.

5. What is the difference between using a number as a subscript and using a number as a coefficient in an equation?

6. Sodium hydrogen carbonate, $NaHCO_3$, is prepared commercially by the following series of reactions:

   1. Calcium carbonate is heated to produce calcium oxide and carbon dioxide.

   2. Ammonium chloride is heated with calcium oxide to produce ammonia, calcium chloride, and water.

   3. Carbon dioxide, ammonia, and water react to yield ammonium hydrogen carbonate.

   4. Ammonium hydrogen carbonate and sodium chloride react to produce sodium hydrogen carbonate and ammonium chloride.

      a) Write equations to represent each step.
      b) Which process is a combination reaction?
      c) Which step is a decomposition reaction?

CHAPTER 9 TEST

True — False

_____ 1.   Stoichiometry is the section of chemistry involving calculations based on mass and mole realtionships of substances in chemical reactions.

_____ 2.   In a limiting reactant problem, you determine which reactant has the fewest moles available.

_____ 3.   The maximum amount of product that can be produced according to the equation, from the amounts of reactants supplied, is the theoretical yield.

_____ 4   A mole ratio is a ratio of the moles of products to the moles of reactants.

_____ 5.   A mole ratio is used to convert the moles of starting substance to the moles of desired substance.

_____ 6.   The limiting reactant of a chemical equation limits the amount of product that can be formed.

_____ 7.   The equation for a reaction should be balanced before doing stoichiometric calculations.

_____ 8.   The mole ratio for converting moles of CO_2 to moles of O_2 in the reaction
$2C_2H_6 + 7 O_2 \rightarrow 4 CO_2 + 6H_2O$ is 4 mol CO_2/7 mol O_2.

_____ 9.   The mole ratio for converting moles of CH_4 to moles of H_2O in the reaction
$2CH_4 + 3 O_2 + 2 NH_3 \rightarrow 2 HCN + 6 H_2O$ is 6 mol H_2O/2 mol CH_4.

_____ 10.   The percentage yield of a product is $\underline{\text{actual yield}}$ x 100.
theoretical yield

_____ 11.   When sucrose ($C_{12}H_{22}O_{11}$) is decomposed to give carbon and water, the number of sugar molecules that react will always be equal to the number of carbon atoms that form.

Multiple Choice (choose the best answer)

_____ 1.   20.0 g of Na_2CO_3 is how many moles?

a) 1.89 mol   b) 2.12 x 10^3 mol   c) 212 mol   d) 0.189 mol

_____ 2.   What is the mass in grams of 0.30 mol of $BaSO_4$?

a) 7.0 x 10^3 g   b) 0.13 g   c) 70. g   d) 700.20 g

_____ 3. How many molecules are in 5.8 g of acetone, C_3H_6O?

    a) 0.10 molecules   b) 6.0×10^{22} molecules
    c) 3.5×10^{24} molecules   d) 6.022×10^{23} molecules

_____ 4. In the reaction between CH_4 and O_2, if 18.0 g of CO_2 are produced, how many grams of H_2O are produced?

    a) 7.36 g   b) 3.68 g   c) 9.0 g   d) 14.7 g

_____ 5. How many moles of CO_2 can be produced by the reaction 5.0 mol of C_2H_4 and 12.0 mol of O_2?

    a) 4.0 mol   b) 5.0 mol   c) 8.0 mol   d) 10.0 mol

_____ 6. How many moles of CO_2 can be produced by the reaction of 0.480 mol of C_2H_4 and 1.08 mol of O_2?

    a) 0.240 mol   b) 0.960 mol   c) 0.720 mol   d) 0.864 mol

_____ 7. How many grams of CO_2 could be produced from 2.0 g of C_2H_4 and 5.0 of O_2?

    a) 5.5 g   b) 4.6 g   c) 7.6 g   d) 6.3 g

Problems 8, 9, and 10 refer to the equation:

$$H_3PO_4 + MgCO_3 \rightarrow Mg_3(PO_4)_2 + CO_2 + H_2O$$

_____ 8. The sequence of coefficients for the balanced equation is

    a) 2, 3, 1, 3, 3   b) 3, 1, 3, 2, 3   c) 2, 2, 1, 2, 3   d) 2, 3, 1, 3, 2

_____ 9. If 20.0 g of carbon dioxide is produced, the number of moles of magnesium carbonate used is

    a) 0.228 mol   b) 1.37 mol   c) 0.910 mol   d) 0.455 mol

_____ 10. If 50.0 g of magnesium carbonate react completely with H_3PO_4, the number of grams of carbon dioxide produced is

    a) 52.2 g   b) 21.6 g   c) 13.1 g   d) 50.0 g

_____ 11. In the reaction $C(s) + O_2(g) \rightarrow CO_2(g)$, 24 g of carbon are mixed with 48 grams of oxygen. The reaction goes to completion. When the chemical reaction comes to an end,

    a) solid carbon will be left over.
    b) oxygen molecules will be left over.
    c) the amount of carbon dioxide formed will be equal in grams to the mass of oxygen gas consumed.
    b) the mass of carbon that reacts will be equal to the mass of the oxygen gas that is consumed.

Reasoning and Expression (use a separate sheet of paper)

1. Nitrogen dioxide can form dinitrogen tetroxide under certain circumstances. Write a balanced equation to show this. How many grams of the reactant are needed to produce a kilogram of the product?

2. The NO that is emitted from factory smokestacks can be removed if it reacts with ammonia to form nitrogen gas and water. Write a balanced equation. Determine the mole ratio:

   a) NO to H_2O b) N_2 to H_2O c) NH_3 to H_2O

3. For the reaction $CS_2 + 3O_2 \rightarrow CO_2 + 2 SO_2$, how many grams of SO_2 will result when 152 g of CS_2 and 48 g of O_2 react?

4. A chemistry student weighs a copper wire and then heats it in a flame. After awhile, she determines that the copper wire is now heavier. Explain considering the fact mass is conserved in chemical reactions.

5. When sugar ferments, ethyl alcohol is produced according to this process:
   $$C_6H_{12}O_6 \rightarrow C_2H_5OH + CO_2$$

   a) Balance the equation.
   b) What mass of alcohol can be obtained from 500. g of sugar?
   c) How many liters of alcohol will result from 2.00 kg of sugar reacting? The density of alcohol is 0.79 g/mL.

6. When limestone (calcium carbonate) is heated, it releases carbon dioxide. The other product in the reaction is lime (calocium oxide).

   a) Write and balance the equation for the process.
   b) How many grams of limestone, whose purity is 95%, are required to yield 250.0 g of lime?

7. If 19.0 g of cobalt are mixed with 19.0 g of $FeCl_2$, cobalt (II) chloride and iron are formed.

   a) Write and balance the equation for the reaction.
   b) How many grams of which reactant will be left over?
   c) How many grams of each product will result?

CHAPTER 10 TEST

True — False

_____ 1. An atom of nitrogen has two elections in a 1s orbital, two electrons in a 2s orbital, and one electron in each of three different 2p orbitals.

_____ 2. The Lewis-dot symbol for calcium is Ca.

_____ 3. In the ground state, electrons tend to occupy orbitals of the lowest possible energy.

_____ 4. The second principal energy level contains only s and p sublevels.

_____ 5. The first f sublevel electrons are in the fourth principal energy level.

_____ 6. The 4s energy sublevel fills before the 3d sublevel because the 4s is at a lower energy level.

_____ 7. In a Lewis-dot representation of an element, the symbol of the element is shown surrounded by the number of electrons in the outermost energy level of the atom.

_____ 8. The element represented by $1s^2 2s^2 2p^6 3s^2 3p^4$ has an atomic number of 16.

_____ 9. An atom of $_{16}S$ has nine sublevel orbitals that contain one or more electrons.

_____ 10. An atom of an element of atomic number 21 has electrons in 11 orbitals.

_____ 11. As the wavelength of a light wave increases, its frequency and energy both decrease.

Multiple Choice (choose the best answer)

_____ 1. The concept of electrons existing in specific orbits around the nucleus was the contribution of

a) Thomson   b) Rutherford   c) Bohr   d) Schrödinger

_____ 2. The correct electron structure for a flourine atom, F, is

a) $1s^2 2s^2 2p^5$ b) $1s^2 2s^2 2p^3 3s^2 3p^1$ c) $1s^2 2s^2 2p^4 3s^1$ d) $1s^2 2s^2 2p^3$

_____ 3. The correct electron structure for $_{48}Cd$ is

a) $1s^2 2s^2 2p^6 3s^2 3p^6 4s^2 3d^{10}$ b) $1s^2 2s^2 2p^6 3s^2 3p^6 4s^2 3d^{10} 4p^6 5s^2 4d^{10}$
c) $1s^2 2s^2 2p^6 3s^2 3p^6 4s^2 3d^{10} 4p^6 4d^4$ d) $1s^2 2s^2 2p^6 3s^2 3p^6 4s^2 4p^6 4d^{10} 5s^2 5d^{10}$

_____ **4.** The correct electron structure of $_{23}$V is

    a) $[Ar]3s^23d^2$ b) $[Ar]4s^24p^3$ c) $[Ar]4s^24d^3$ d) $[Ar]4s^23d^3$

_____ **5.** Which of the following is the correct atomic structure for $_{22}^{48}$Ti?

    a) (22p 26n) 2 8 10 2   b) (26p 22n) 2 8 8 4

    c) (22p 48n) 2 8 8 4   d) (22p 26n) 2 8 8 4

_____ **6.** The number of orbitals in a d sublevel is

    a) 3   b) 5   c) 7   d) No correct answer given.

_____ **7.** The number of electrons in the third principal energy level in an atom having the electron structure $1s^22s^22p^63s^23p^2$ is

    a) 2   b) 4   c) 6   d) 8

_____ **8.** Which of these elements has two s and six p electrons in its outer energy level?

    a) He   b) O   c) Ar   d) No correct answer given.

_____ **9.** Which element is *not* a noble gas?

    a) Ra   b) Xe   c) He   d) Ar

_____ **10.** Which of the following elements has the largest number of unpaired electrons?

    a) F   b) S   c) Cu   d) N

_____ **11.** Each of the following statements about the electromagnetic spectrum except which is true?

    a) X-rays have longer wavelengths than microwaves.
    b) Red light is lower in energy than blue.
    c) When a 2s electron becomes a 1s electron, EM radiation is absorbed.
    d) All colors of light travel at the same speed.

_____ **12.** The energy of electromagnetic radiation is greater if

    a) its wavelength is short and its frequency is high.
    b) its wavelength is short and its frequency is low.
    c) its wavelength is long and its frequency is low.
    d) its wavelength is long and its frequency is high.

Reasoning and Expression (use a separate sheet of paper)

1. Draw a sketch of a wave showing two wavelengths. Superimpose another wave on top of the first, the second wave having half the amplitude and twice the wavelength of the first.

2. From the periodic table, write the symbol for all the elements which

   a) have no p electrons   b) have two to four d electrons   c) have exactly half-filled p sublevels

3. What fraction of its total electrons are p electrons in atoms of

   a) Be?   b) Ar?   c) Cs?

4. What is the periodic table relationship between atomic size (radius) and atomic number? Why is there not a significant change in radius from element 17 to 18, but a significant change from element 18 to 19?

5. How many unpaired electrons are in the valence level of

   a) B   b) Se   c) Cl   d) Kr

6. Some types of electromagnetic radiation are not visible to the human eye. What evidence do you have for the existence of:

   a) infrared light
   b) ultraviolet light

CHAPTER 11 TEST

True — False

_____ **1.** A bromide ion is smaller than a bromine atom.

_____ **2.** When elements combine by ionic bonding, they normally form molecules.

_____ **3.** Oxygen has a greater electronegativity than carbon.

_____ **4.** The formula of the aluminum ion is Al^+.

_____ **5.** A phosphate ion, PO_4^{3-}, is a polyatomic ion.

_____ **6.** The Lewis-dot symbol for $_{84}PO$ is $\cdot\ddot{P}o\colon$

_____ **7.** The ionic bond is the electrostatic attraction existing between oppositely charged ions.

_____ **8.** Two atoms with different electronegativities that form a polar covalent bond will have partial ionic charges.

_____ **9.** The covalent bond between two carbon atoms will be nonpolar.

_____ **10.** A covalent bond formed by sharing one pair of electrons is called a single bond.

_____ **11.** Ionic compounds consist of charged particles bonded together in a crystal lattice.

_____ **12.** As the wavelength of a light wave increases, its frequency and energy both decrease.

Multiple Choice (choose the best answer)

_____ **1.** Which of the following molecules is a dipole?

a) HBr   b) CH_4 c) H_2 d) CO_2

_____ **2.** Which of the following is a correct Lewis structure?

a) $:\ddot{O}:C:\ddot{O}:$ b) $:\ddot{C}l:\ddot{C}:\ddot{C}l:$ with $:\ddot{C}l:$ above and $:\ddot{C}l:$ below   c) $\ddot{C}l::\ddot{C}l$ d) $:\ddot{N}:\ddot{N}:$

_____ **3.** The correct Lewis structure for SO_2 is

a) :Ö:S:Ö: **b)** :Ö:S::Ö: **c)** :Ö::S::Ö: **d)** :Ö:S:Ö:

_____ **4.** Carbon dioxide, CO_2, is a nonpolar molecule because

a) Oxygen is more electronegative than carbon.
b) The two oxygen atoms are bonded to the carbon atom.
c) The molecule has a linear structure with the carbon atom in the middle.
d) The carbon-oxygen bonds are polar covalent.

_____ **5.** When a magnesium atom participates in a chemical reaction, it is most likely to

a) lose 1 electron   b) gain 1 electron   c) lose 2 electrons   d) gain 2 electrons

_____ **6.** If X represents an element of Group IIIA, what is the general formula for its oxide?

a) X_3O_4 b) X_3O_2 c) XO d) X_2O_3

_____ **7.** Which compound forms a tetrahedral molecule?

a) $NaCl$ b) CO_2 c) CH_4 d) $MgCl_2$

_____ **8.** Which compound has double bonds within its molecular structure?

a) $NaCl$ b) CO_2 c) CH_4 d) H_2O

_____ **9.** The number of electrons in a triple bond is

a) 3   b) 4   c) 6   d) 8

_____ **10.** Which of the following does *not* have a noble-gas electron structure?

a) Na b) Sc^{3+} c) Ar d) O^{2-}

_____ **11.** Which of these characterizes covalent bonding?

a) The formation of ions
b) The type of bond usually found between metals and nonmetals
c) The loss of valence electrons
d) The formation of true, discrete molecules

_____ **12.** Atom P has 2 valence electrons and atom Q has 5 valence electrons. The formula expected for the ionic compound that forms between P and Q is

a) P_3Q_2
b) P_2Q_5
c) P_2Q_3
d) PQ_3

_____ **13.** Which series is ranked in order of increasing electronegativity?

    a) O, S, Se, Te
    b) Cl, S, P, Si
    c) In, Sn, N, O
    d) C, Si, P, Se

Reasoning and Expression (use a separate sheet of paper)

1. Arrange the following bonds in order of increasing polarity.

    a) N—O   b) Br—Cl   c) H—C   d) Li—F   e) I—I

2. When BH_3 reacts with H^-, the product is BH_4^-. Identify what happens to the shape of the molecule as the reaction occurs. Draw Lewis structures to support your position.

3. Draw Lewis structures for

    a) CO_3^{2-} b) N_3^- c) $COCl_2$

4. What do the Lewis structures of OH, CH_3 and NO_2 have in common? What does this mean for the stability of those species?

5. Consider $MgCl_2$ and SCl_2. Compare and contrast these two compounds with respect to

    a) Lewis structures   b) types and number of bonds involved
    c) shape of the molecules   d) polarity of the molecules

6. If a molecule consists of a central atom surrounded by three other atoms, why isn't it necessarily trigonal planar in shape?

CHAPTER 12 TEST

True — False

_____ 1. Pressure is defined as force per unit area.

_____ 2. One torr is equal to 760 mm Hg.

_____ 3. Boyle's law states that the volume of a gas is inversely proportional to the pressure.

_____ 4. Charles' law states that the pressure of a gas is directly proportional to the temperature.

_____ 5. Avogadro's law states that equal volumes of different gases at the same temperature and pressure contain the same masses of gas.

_____ 6. An ideal gas is one whose behavior is described exactly by the gas laws for all possible values of P, V, and T.

_____ 7. According to Avogadro, there are 6.022×10^{23} molecules of gas in a liter at STP.

_____ 8. At constant temperature and pressure, N_2 will effuse more rapidly than O_2.

_____ 9. The volume of a gas is dependent on the number of gas molecules, the pressure, the absolute temperature, and the gas constant R.

_____ 10. Ozone in the stratosphere aids ultraviolet radiation in reaching the earth's surface.

_____ 11. When ideal gas molecules collide, the total KE of those molecules greatly increases.

Multiple Choice (choose the best answer)

_____ 1. The ratio of the relative rate of effusion of methane, CH_4, to sulfur dioxide, SO_2, is

a) $\dfrac{64}{16}$ b) $\dfrac{16}{64}$ c) $\dfrac{1}{4}$ d) $\dfrac{2}{1}$

_____ 2. Measured at 65°C and 500. torr, the mass of 3.21 L of a gas is 3.5 g. The molar mass of this gas is

a) 21 g/mole   b) 46 g/mole   c) 24 g/mole   d) 130 g/mole

_____ 3. A 300. mL sample of oxygen, O_2, was collected over water at 23°C and 725 torr pressure. If the vapor pressure of water at 23°C is 21.0 torr, the volume of dry O_2 at STP would be

a) 256 mL   b) 351 mL   c) 341 mL   d) 264 mL

_____ **4.** A tank containing 0.01 mol of neon and 0.04 mol of helium shows a pressure of 1 atm. What is the partial pressure of neon in the tank?

a) 0.8 atm   b) 0.01 atm   c) 0.2 atm   4) 0.5 atm

_____ **5.** How many liters of NO_2 (at STP) can be produced from 25.0 g of Cu reacting with concentrated nitric acid?

$$Cu(s) + 4\,HNO_3(aq) \rightarrow CU(NO_3)_2(aq) + 2\,H_2O(l) + 2\,NO_2(g)$$

a) 4.41 L   b) 8.82 L   c) 17.6 L   d) 44.8 L

_____ **6.** Which of the following is *not* one of the principal assumptions of the Kinetic-Molecular Theory for an ideal gas?

a) All collisions of gaseous molecules are perfectly elastic.
b) A mole of any gas occupies 22.4 L at STP.
c) Gas molecules have no attraction for one another.
d) The average kinetic energy for molecules is the same for all gases at the same temperature.

_____ **7.** If the pressure on 45 mL of gas is changed from 600. torr to 800. torr, the new volume will be

a) 60 mL   b) 34 mL   c) 0.045 mL   d) 22.4 L

_____ **8.** The volume of a gas is 300. mL at 740. torr and 25°C. If the pressure remains constant and the temperature is raised to 100.°C, the new volume will be

a) 240. mL   b) 1.20 L   c) 376 mL   d) 75.0 mL

_____ **9.** Which of these gases has the highest density at STP?

a) N_2O b) NO_2 c) Cl_2 d) SO_2

_____ **10.** What is the density of CO_2 at 25°C and 0.954 atm pressure?

a) 1.72 g/L   b) 2.04 g/L   c) 0.985 g/L   d) 1.52 g/L

_____ **11.** Which of the graphs below represents the relationship between temperature and pressure of an ideal gas, if volume and moles are held constant

Reasoning and Expression (**use a separate sheet of paper**)

1. What is the density of sulfur hexafluoride at room temperature and 680 mm Hg?

2. Sketch graphs to indicate the relationship between

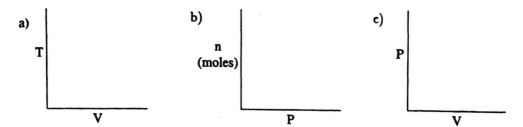

3. Use the Kinetic-Molecular Theory to explain why

   a) bread dough rises in the oven
   b) you feel cool after getting out of a swimming pool, even on a warm day

4. A small 0.500 mL bubble forms near the bottom of a lake where temperature is 2°C and the pressure is 2.40 atmospheres. What volume will the bubble occupy near the surface where the temperature is now 18°C and the pressure is 770. mm?

5. Just 0.020 g of an unknown gas is contained in a 250 ml tube at a temperature of 305 K and a pressure of 4.6 kPa. What is its molar mass?

6. Here is a diagram of a sample of an ideal gas at 30°C and 0.5 atm.

   Which of these diagrams shows the same gas at STP? Explain your choice briefly.

a)   b)   c)   d)   e)

CHAPTER 13 TEST

True — False

_____ 1.   Calcium oxide reacts with water to form calcium hydroxide and hydrogen gas.

_____ 2.   At pressures below 760 torr, water will boil above 100°C.

_____ 3.   Evaporation is the escape of molecules from the liquid state to the vapor state.

_____ 4.   Sublimation is the change from the vapor to the liquid state.

_____ 5. $CuSO_4 \cdot 5H_2O$ is named copper(II) sulfate pentahydrate.

_____ 6.   Hard water contains relatively large amounts of Ca^{2+} and Mg^{2+} ions.

_____ 7.   Substances that evaporate readily are said to be volatile.

_____ 8. SO_2 is the anhydride of H_2SO_4.

_____ 9.   Ice at 0°C is at the same temperature as water at 0°C, but ice contains less heat energy than the water.

_____ 10.   The vapor pressure of a liquid depends on the temperature and the atmospheric pressure.

_____ 11.   Hydrogen bonding occurs within molecules, and covalent bonding occurs between molecules.

Multiple Choice (choose the best answer)

_____ 1.   The heat of fusion of water is

   a) 4.184 J/g   b) 335 J/g   c) 2.26 kJ/g   d) 2.26 kJ/mol

_____ 2.   The heat of vaporization of water is

   a) 4.184 J/g   b) 335 J/g   c) 2.26 kJ/g   d) 2.26 kJ/mol

_____ 3. SO_2 can be properly classified as a (n)

   a) basic anhydride   b) hydrate   c) anhydrous salt   d) acid anhydride

_____ 4. When compared to H_2S, H_2Se, and H_2Te, water is found to have the highest boiling point because it

    a) has the lowest molar mass   b) is the smallest molecule
    c) has the highest bonding   d) will form hydrogen bonds better than the others

_____ 5. Which of the following is an incorrect equation?

    a) $H_2SO_4 + 2\, NaOH \rightarrow Na_2SO_4 + 2\, H_2O$ b) $C_2H_6 + O_2 \rightarrow 2\, CO_2 + 3\, H_2$

    c) $2\, H_2O \xrightarrow[\text{H}_2\text{SO}_4]{\text{Electrolysis}} 2\, H_2 + O_2$ d) $Ca + 2\, H_2O \rightarrow H_2 + Ca(OH)_2$

_____ 6. How many kilojoules are required to change 85 g of water at 25°C to steam at 100.°C?

    a) 219 kJ   b) 27 kJ   c) 590 kJ   d) 192 kJ

_____ 7. A chunk of 0°C ice, mass 145 g, is dropped into 75 g of water at 62°C. The heat of fusion of water is 335 J/g. The result, after thermal equilibrium is attained, will be

    a) 87 g of ice and 133 g of liquid water, all at 0°C
    b) 58 g of ice and 162 g of liquid water, all at 0°C
    c) 220 g of water at 7°C
    d) 220 g of water at 17°C

_____ 8. Hydrogen bonding

    a) occurs only between water molecules
    b) is stronger than covalent bonding
    c) can occur between NH_3 and H_2O
    d) results from strong attractive forces in ionic compounds

_____ 9. A liquid boils when

    a) the vapor pressure of the liquid equals the external pressure above the liquid
    b) the heat of vaporization exceeds the vapor pressure
    c) the vapor pressure equals one atmosphere
    d) the normal boiling temperature is reached

_____ 10. 95.0 g of 0.0°C ice is added to exactly 100. g of water at 60.0°C. When the temperature of the mixture first reaches 0.0°C, the mass of ice still present is

    a) 0.0 g   b) 20.0 g   c) 10.0 g   d) 75.0 g

_____ 11. The vapor pressure of a liquid depends on

    a) temperature and intermolecular forces
    b) boiling point and intermolecular forces
    c) temperature and boiling point
    d) temperature, boiling point, intermolecular forces, and volume of the liquid

Reasoning and Expression (use a separate sheet of paper)

1. For a certain substance, the normal boiling point is 85°C, the normal freezing point is 15°C. Its heat of fusion is 20 kJ/g and its heat of vaporization is 45 kJ/g.

    a) Sketch a heating curve for this substance.
    b) Determine the heat needed to boil 125 g of it.
    c) Briefly describe what you would expect to see as some of this substance is warmed from 10°C to 100°C.

2. Differentiate between

    a) melting point and boiling point
    b) heat of fusion and freezing point
    c) boiling and subliming

3. How many grams of ice at 0°C can be melted and completely converted to steam if 1×10^6 J of energy are added?
    (heat of fusion = 340 J/g; heat of vaporization = 2250 J/g; specific heat = 4.2 J/g°C)

4. How would you anticipate the shape of the cooling curve of a material would change or differ if you warmed 10 g of the substance or if you warmed 1 kg of it? Explain.

5. Complete and balance the following equations

    a) $SO_2 + H_2O \rightarrow$ b) $Li_2O + H_2O \rightarrow$ c) $Na + H_2O \rightarrow$

6. Here are two molecules of acetic acid, the characteristic component of vinegar.

    ```
      H  O              H  O
      |  ||             |  ||
    H—C—C—O—H         H—C—C—O—H
      |                 |
      H                 H
    ```

    a) Draw the hydrogen bond(s) between them.
    b) Why do liquid molecules that engage in hydrogen bonding tend to have higher boiling points?

CHAPTER 14 TEST

True — False

_____ 1. Bromine is more soluble in polar water than in nonpolar carbon tetrachloride because the polar water molecules help it form ions.

_____ 2. An increase in temperature almost always increases the rate at which a solid will dissolve in a liquid.

_____ 3. Freezing point depression, boiling point elevation, and vapor pressure lowering are colligative properties of solutions.

_____ 4. Liquids that mix with water in all proportions are usually ionic in solution or are polar substances.

_____ 5. Salt water has a higher boiling point than distilled water.

_____ 6. One mole of H_2SO_4 will react with twice as much NaOH as 1 mol of HCl.

_____ 7. Large crystals dissolve faster than do smaller ones because they expose a larger surface area to the solvent.

_____ 8. The molarity of a solution is the number of moles of solute per liter of solution.

_____ 9. A solution containing 0.001 mol of solute in 1 mL of solvent is 1 M.

_____ 10. About 42 g of NaCl are dissolved in 100 mL of water in a glass at room temperature. As more of the solute is added, some NaCl sinks to the bottom of the glass and collects there. This solution is correctly described as supersaturated.

Multiple Choice (choose the best answer)

_____ 1. Which of the following is *not* a general property of solutions?

    a) It is a homogeneous mixture of two or more substances.
    b) It has a variable composition.
    c) The dissolved solute breaks down to individual molecules.
    d) The solution has the same chemical composition, the same chemical properties, and the same physical properties in every part.

_____ **2.** If 5.00 g of NaCl are dissolved in 25.0 g of water, the percent of NaCl by mass is

a) 16.7 b) 20.0 c) 0.20 d) No correct answer given.

_____ **3.** How many grams of 9.0% $AgNO_3$ solution will contain 5.3 g $AgNO_3$?

a) 47.7 b) 0.58 c) 59 d) No correct answer given.

_____ **4.** If 400. mL of 2.0 M HCl react with excess $CaCO_3$, the volume of CO_2 produced, measured at STP, is

a) 18 L b) 5.6 L c) 9.0 L d) 56 L

_____ **5.** If 20.0 g of the nonelectrolyte urea, $CO(NH_2)_2$, are dissolved in 25.0 g of water, the freezing point of the solution will be

a) -2.48°C b) -1.40°C c) -24.8°C d) -3.72°C

_____ **6.** When 256 g of a nonvolatile, nonelectrolyte unknown were dissolved in 500. g of H_2O, the freezing point was found to be -2.79°C. The molar mass of the unknown solute is

a) 357 g/mol b) 62.0 g/mol c) 768 g/mol d) 341 g/mol

_____ **7.** Which procedure is least likely to increase the solubility of most solids in liquids?

a) Stirring
b) Pulverizing the solid
c) Heating the solution
d) Increasing the pressure

_____ **8.** The addition of a crystal of $NaClO_3$ to a solution of $NaClO_3$ causes additional crystals to precipitate. The original solution was

a) unsaturated b) dilute c) saturated d) supersaturated

_____ **9.** A solution of ethyl alcohol and benzene is 40% alcohol by volume. Which statement is false?

a) The solution contains 40 mL of alcohol in 100 mL of solution.
b) The solution contains 60 mL of benzene in 100 mL of solution.
c) The solution contains 40 mL of alcohol in 100 g of solution.
d) The solution is made by dissolving 40mL of alcohol in 60 mL of benzene.

_____ **10.** To what volume must 10.0 mL of 10.0 M HNO_3 be diluted to produce a 0.100 M solution?

a) 1.00 L b) 500 mL c) 2.00 mL d) 1.50 L

_____ **11.** If you have a solution containing silver ions, the addition of each of the following would cause a precipitate to form *except*

a) Cl^- b) CO_3^{2-} c) NO_3^- d) Br^-

Reasoning and Expression (use a separate sheet of paper)

1. Three solutions are mixed together: 100ml of 0.05 M $AgNO_3$, 250 ml of 1.5 M $AgNO_3$ and 650 ml of 0.45 M $AgNO_3$. What volume of what concentration do you have in the final mixture?

2. An average cup (250 ml) of coffee contains 125 mg of caffeine, $C_8H_{10}N_4O_2$. What is the molarity of caffeine in the coffee? What mass percent is this?

3. Briefly describe what you predict the effect of increased temperature would be on

   a) the solubility of a solid in a liquid
   b) the solubility of a gas in a liquid

4. How many grams of ethyl alcohol (C_2H_5OH) are needed to drop the freezing temperature of 2.0 L of water to $-10.°C$? How many grams of sugar ($C_{12}H_{22}O_{11}$) would be needed to produce the same effect?

5. Suppose you wish to make a 1 M solution of salt (NaCl) and a 1 M solution of sugar ($C_{12}H_{22}O_{11}$). Describe in words how you would do each.

6. How can a solution be accurately described as both dilute and saturated? What kind of solute would it have to contain?

Name ______________________
Section __________ Date __________
Instructor ______________________

CHAPTER 15 TEST

True — False

_____ **1.** Arrhenius defined a base as a hydroxide-containing substance that dissociates in water to produce hydroxide ions.

_____ **2.** The Bronsted-Lowry theory defined an acid as a proton donor.

_____ **3.** In the reaction $HNO_3 + H_2O \rightarrow H_3O^+ + NO_3^-$ the conjugate base of HNO_3 is NO_3^-.

_____ **4.** When 1 mol of $CaCl_2$ dissolves in water, it will give 1 mol of Ca^{2+} and 2 mol of Cl^- ions in solution.

_____ **5.** In neutralizations between a strong acid and a strong base, the net ionic equation is $H^+ + OH^- \rightarrow H_2O$.

_____ **6.** In writing net ionic equations, weak electrolytes are written in their un-ionized form.

_____ **7.** Negative ions are called anions.

_____ **8.** When 50.0 mL of 0.20 M NaOH and 100. mL of 0.10 M HCl are mixed, the resulting solution will have a pH of 7.

_____ **9.** A 0.1 M HCI and a 0.1 M $HC_2H_3O_2$ solution will have the same pH.

_____ **10.** When water ionizes at 100°C it produces more H^+ ions than OH^- ions.

_____ **11.** A 1 M solution of a strong base will have both a greater conductivity and a higher pH than a 1 M solution of a weak base.

_____ **12.** The compound C_2H_5OH is a strong electrolyte in aqueous solution.

***Multiple Choice* (choose the best answer)**

_____ **1.** When the reaction $Al + HCl \rightarrow$ is completed and balanced, a term appearing in the balanced equation is

a) 3 HCl b) $AlCl_2$ c) 3 H_2 d) 4 Al

_____ **2.** When the reaction $HCl + Cr_2(CO_3)_3 \longrightarrow$ is completed and balanced, a term appearing in the balanced equation is

a) Cr_2Cl b) $3\,HCl$ c) $3\,CO_2$ d) H_2O

_____ **3.** Which of the following is a nonelectrolyte?

a) $HC_2H_3O_2$ b) $MgSO_4$ c) $KMnO_4$ d) CCl_4

_____ **4.** Which of the following is a strong electrolyte?

a) H_2CO_3 b) HNO_3 c) NH_4OH d) H_3BO_3

_____ **5.** A solution has a concentration of H^+ of 3.4×10^{-5} M. The pH is

a) 4.47   b) 5.53   c) 3.53   d) 5.47

_____ **6.** 16.55 mL of 0.844 M NaOH is required to titrate 10.00 mL of a hydrochloric acid solution. The molarity of the acid solution is

a) 0.700 M   b) 0.510 M   c) 1.40 M   d) 2.55 M

_____ **7.** What volume of 0.462 M NaOH is required to neutralize 20.00 mL of 0.391 M HNO_3?

a) 23.6 mL   b) 16.9 mL   c) 9.03 mL   d) 11.8 mL

_____ **8.** 25.00 mL of H_2SO_4 solution required 18.92 mL of 0.1024 M NaOH for complete neutralization. The normality of the acid is

a) 0.1550 M   b) 0.07750 M   c) 0.03875 M   d) 0.6765 M

_____ **9.** Dilute hydrochloric acid is a typical acid, as shown by its

a) color   b) odor   c) solubility   d) taste

_____ **10.** The amount of $BaSO_4$ that will precipitate when 100. mL of 0. 10 M $BaCl_2$ and 100. mL of 0.10 M Na_2SO_4 are mixed is

a) 0.010 mol   b) 0.10 mol   c) 23 g   d) No correct answer given.

_____ **11.** Which of the following processes would *not* allow one to clearly distinguish an acidic solution from a basic one?

a) By adding a piece of Mg to the solution to see if hydrogen gas results.
b) By determining the value of the solution with a pH meter.
c) By testing the conductivity of the solution.
d) By noting the color the solution turns in the presence of an indicator.

_____ **12.** 5.6 grams of a weak monoprotic acid, HA, was dissolved in water. Titration of the acid to its end point required 58.4 mL of a 0.50 M KOH solution. What is the molar mass of the acid, HA?

 a) 192 g/mol   b) 164 g/mol
 c) 96.0 g/mol   d) 5.21 g/mol

***Reasoning and Expression* (use a separate sheet of paper)**

1. Complete the following Bronsted-Lowry equations. In each case, label the conjugate acid

 a) $HSO_4^- + H_2O \rightarrow$ b) $CO_3^{2-} + NH_4^+ \rightarrow$ c) $HS^- + NO_2^- \rightarrow$ d) $OH^- + H_2O \rightarrow$

2. Two hundred milliliters of 0.50 M HCl are diluted to 5 liters with water.

 a) What is its molarity?
 b) What will be its pH?
 c) How many grams of NaOH would be needed to completely neutralize this solution?

3. Ammonia, NH_3, is identified as a base by the Bronsted definition, but would not be described as such by the Arrhenius definition. Explain, using equations.

4. When Na_2CO_3 is mixed with $Ca(OH)_2$, a precipitate of $CaCO_3$ forms. If 100 mL of 1.00 M Na_2CO_3 are mixed with 200 mL 2M $Ca(OH)_2$, how many grams of $CaCO_3$ will precipitate?

5. If 75.0 g of aluminum dichromate [$Al_2(Cr_2O_7)_3$] is dissolved to make 2.00 L of solution, what is the the molarity of each ion in the solution? Write an equation to show the dissociation of the compound in water.

6. Suppose that 50.0 mL of 1 M NaOH are mixed with 49.5 mL of 1 M HI.
 a) Is this solution neutral?
 b) Will phenolphthalein turn pink when it is added to the solution?
 c) What is the pH of the mixture?
 d) What volume of which solution is needed to reach a pH of 7?

CHAPTER 16 TEST

True — False

_____ 1. A reversible reaction is one in which the products formed in a chemical reaction are reacting to produce the original reactants.

_____ 2. A statement of Le Chatelier's principle is that, if stress is applied to a system in equilibrium the system will behave in such a way as to relieve that stress and restore equilibrium but under a new set of conditions.

_____ 3. A catalyst will shift the point of equilibrium of a reaction but will not alter the reaction rates.

_____ 4. If K_a for acetic acid is 1.8×10^{-5}, and K_a for nitrous acid is 4.5×10^{-4}, then, at equal concentrations, acetic acid is a stronger acid.

_____ 5. The amount of energy needed to form an activated complex is known as the activation energy of a reaction.

_____ 6. A chemical reaction at equilibrium will have different equilibrium constants at different temperatures.

_____ 7. When the temperature of an exothermic reaction is increased, the forward reaction is favored.

_____ 8. A solution made from NaCl and HCl will act as a buffer solution.

_____ 9. An increase in pressure for a system composed entirely of gases will cause the reaction to shift to the side that contains the larger number of moles.

_____ 10. The K_{sp} expression for $Fe(OH)_3(s) \rightleftarrows Fe^{3+} + 3(OH)^-$ is: $K_{sp} = \dfrac{[Fe^{3+}][OH^-]^3}{[Fe(OH)_3]}$

_____ 11. The reaction equation $BaCO_3(s) \rightleftarrows Ba^{2+} + CO_3^{2-}$ represents an acid/base equilibrium system.

Multiple Choice (choose the best answer)

_____ 1. The equation $HC_2H_3O_2 + H_2O \rightleftarrows H_3O^+ + C_2H_3O_2^-$ implies that

a) If you start with 1.0 mol of $HC_2H_3O_2$, 1.0 mol of H_3O^+ and 1.0 mole of $C_2H_3O_2^-$ will be produced.
b) An equilibrium exists between the forward reaction and the reverse reaction.
c) At equilibrium, equal molar amounts of all four substances will exist.
d) The reaction proceeds all the way to the products, then reverses, going all the way back to the reactants.

_____ 2. If the reaction $A + B \rightleftarrows C + D$ is initially at equilibrium, and then more A is added, which of the following is not true?

a) More collisions of A and B will occur, thus the rate of the forward reaction will be increased.
b) The equilibrium will shift to the right.
c) The moles of B will be increased.
d) The moles of D will be increased.

_____ 3. If $[H^+] = 2.0 \times 10^{-4}$ M then [OH-] will be

a) 5.0×10^{-9} M   b) 3.70   c) 2.0×10^{-4} M   d) 5.0×10^{-11} M

_____ 4. Which of the following solutions would be the best buffer solution?

a) 0.10 M $HC_2H_3O_2$ + 0.10 M $NaC_2H_3O_2$ b) 0.10 M HCl
c) 0.10 M HCl + 0.10 M NaCl   d) Pure water

_____ 5. The equilibrium constant for the reaction $2 A + B \rightleftarrows 3 C + D$ is

a) $\dfrac{[C]^3[D]}{[A]^2[B]}$ b) $\dfrac{[2A][B]}{[3C][D]}$ c) $\dfrac{[3C][D]}{[2A][B]}$ d) $\dfrac{[A]^2[B]}{[C]^3[D]}$

_____ 6. Which factor will *not* increase the concentration of ammonia as represented by the following equation?

$$3 H_{2\,(g)} + N_{2\,(g)} \rightleftarrows 2 NH_{3\,(g)} + 92.5 \text{ kJ}$$

a) Increasing the temperature
b) Increasing the concentration of N_2
c) Increasing the concentration of H_2
d) Increasing the pressure

_____ 7. If HCl (g) is added to a saturated solution of AgCl, the concentration of Ag^+ in solution

a) increases   b) decreases
c) remains the same   d) increases and decreases irregularly

_____ **8.** The K_{sp} for $BaCrO_4$ is 8.5×10^{-11}. What is the solubility of $BaCrO_4$ in grams per liter?

   a) 9.2×10^{-6} b) 0.073   c) 2.3×10^{-3} d) 8.5×10^{-11}

_____ **9.** Which would occur if a small amount of sodium acetate crystals, $NaC_2H_3O_2$, were added to 100 mL of 0.1 M $HC_2H_3O_2$ at constant temperature?

   a) The number of acetate ions in the solution would decrease.
   b) The number of acetic acid molecules would decrease.
   c) The number of sodium ions in solution would decrease.
   d) The H^+ concentration in the solution would decrease.

_____ **10.** If the temperature is decreased for the endothermic reaction: $A + B \rightleftarrows C + D$ which of the following is true?

   a) The concentration of A will increase.
   b) No change will occur.
   c) The concentration of B will decrease.
   d) The concentration of D will increase.

_____ **11.** In an equilbrium system at the molecular level,

   a) the chemical reaction has stopped.
   b) the reactant molecules have been completely consumed.
   c) heat must be applied to restart the reaction.
   d) products are being formed just as quickly as reactant molecules are disappearing.

Reasoning and Expression (use a separate sheet of paper)

1. Consider the reaction: heat $+ N_2 {}_{(g)} + 3 H_2 {}_{(g)} \rightleftarrows 2 NH_3 {}_{(g)}$. Identify four changes in the system which would increase the production of ammonia.

2. A 7.5 container at equilibrium contains 10.0 g of NO_2 and 0.55 g of N_2O_4. Calculate the equilibrium constant for this reaction: $2NO_2(g) \rightleftarrows N_2O_4(g)$.

3. What is the percent ionization in a 1.0 M solution of acetic acid? What is it, if the solution of acetic acid is 0.010 M? ($K_a = 1.8 \times 10^{-5}$)

4. The K_{sp} for silver sulfide, Ag_2S, is 5.4×10^{-49}. What is the concentration of each ion in a saturate solution of silver sulfide?

5. Suppose you attempt to dissolve several teaspoons of sugar in a glass of iced tea. Some sugar remains on the bottom of the glass. You add an additional teaspoon of sugar. Will the solution taste any sweeter? Explain why or why not.

6. Calculate the percentage ionization for

   a) 1 M acetic acid
   b) 0.1 M acetic acid
   c) 0.01 M acetic acid

   Comment on your answers.

CHAPTER 17 TEST

True — False

_____ 1. In oxidation, the oxidation number of an element increases in a positive direction as a result of gaining electrons.

_____ 2. The oxidation number of chlorine in Cl_2 is - 1.

_____ 3. The change in the oxidation number of an element from -2 to 0 is reduction.

_____ 4. A free metal can displace from solution the ions of a metal that lies below the free metal in the Activity Series.

_____ 5. In a lead storage battery, $PbSO_4$ is produced at both electrodes in the discharging cycle.

_____ 6. As a lead storage battery discharges, the electrolyte becomes less dense.

_____ 7. The algebraic sum of the oxidation numbers of all the atoms in $K_2Cr_2O_7$ is zero.

_____ 8. A reducing agent will decrease in oxidation number when used in a redox reaction.

_____ 9. In the reaction $2\ Cl^- \rightarrow Cl_2$, each chloride ion loses one electron.

_____ 10. In an electrolytic cell, electrical energy is used to bring about a chemical reaction.

_____ 11. The decomposition of $CaCO_3 \rightarrow CaO + CO_2$ is an example of an oxidation/reduction reaction.

Multiple Choice (choose the best answer)

_____ 1. In K_2SO_4, the oxidation number of sulfur is

a) $+2$ b) $+4$ c) $+6$ d) -2

_____ 2. In $Ba(NO_3)_2$, the oxidation number of N is

a) $+5$ b) -3 c) $+4$ d) -1

_____ 3. Which of the following pairs will *not* react in water solution?

a) $Zn, CuSO_4$ b) $Cu, Al_2(SO_4)_3$ c) $Fe, AgNO_3$ d) $Mg, Al_2(SO_4)_3$

_____ **4.** Which element is the most easily oxidized?

a) K   b) Mg   c) Zn   d) Cu

_____ **5.** Which element will reduce Cu^{2+} to Cu but will _not_ reduce Zn^{2+} to Zn?

a) Fe   b) Ca   c) Ag   d) Mg

_____ **6.** In the electrolysis of fused (molten) $CaCl_2$, the product at the negative electrode is

a) Ca^{2+} b) Cl^- c) Cl_2 d) Ca

_____ **7.** In its reactions, a free element from Group IIA in the periodic table is most likely to

a) be oxidized   b) be reduced   c) be unreactive   d) gain electrons

_____ **8.** Which of the following ions can be reduced by H_2?

a) Hg^{2+} b) Sn^{2+} c) Zn^{2+} d) K^+

_____ **9.** Which of the following is _not_ true of a zinc-mercury cell?

a) It provides current at a steady potential.
b) It has a short service life.
c) It is self-contained.
d) It can be stored for long periods of time.

_____ **10.** Which reaction does not involve oxidation-reduction?

a) Burning sodium in chlorine   b) Chemical union of Fe and S
c) Decomposition of $KClO_3$ d) Neutralization of NaOH with H_2SO_4

_____ **11.** A reduction reaction always

a) requires a loss of electrons
b) requires a simultaneous oxidation process
c) occurs at the anode in a voltaic cell
d) consumes the substances known as the reducing agent

_____ **12.** Which is the most powerful oxidizing agent?

a) iodine   b) sodium   c) chlorine   d) SO_2

_____ **13.** The oxidation number for sulfur in K_2SO_3 is

a) -1   b) +3   c) +4   d) +6

Reasoning and Expression (use a separate sheet of paper)

1. Balance these equations:

   a) $P + HNO_3 \rightarrow HPO_3 + NO + H_2O$
   b) $S^{2-} + Cl_2 \rightarrow SO_4^{2-} + Cl^-$ (basic solution)
   c) $MnO_4^- + Cl^- \rightarrow Mn^{2+} + Cl_2$ (acidic solution)

2. In this galvanic cell, a strip of Zn in $Zn(NO_3)_2$ is paired with a strip of Ag in a solution of $AgNO_3$.

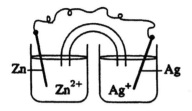

   a) In which direction do electrons flow?
   b) Which metal is oxidized?
   c) In which direction do cations flow?
   d) Write equations for half-reactions that occur
      in each beaker.

3. In this equation, $I_2 + SO_3^{2-} + 2OH^- \rightarrow H_2O + SO_4^{2-} + 2I^-$ identify the oxidizing and reducing agents.

4. Each of these compounds contain P. Arrange them in order of increasing oxidation number:
   H_3PO_3 P_4 P_4O_{10} PF_5 $Ca_3(PO_4)_2$

5. When aluminum foil is placed in a solution of copper (II) ions, a vigorous reaction occurs. Explain in words the electron transfer taking place. Write an equation to show this reaction.

6. In an alkaine battery, these two reactions occur:

   $$Zn(s) + 2\,OH^-\,(aq) \rightarrow Zn(OH)_2(s) + 2e^-$$
   $$MnO_2(s) + H_2O + e^- \rightarrow MnO(OH)(s) + OH^-$$

   Which reaction occurs at the anode? Which at the cathode? What is the overall balanced equation?

CHAPTER 18 TEST

True — False

_____ 1. A gamma ray has more penetrating ability than either an alpha or a beta particle.

_____ 2. In an electromagnetic field, an alpha particle undergoes greater deflection than does a beta particle.

_____ 3. The half-life is the time required for one-half of a specified amount of a radioactive nuclide to disintegrate.

_____ 4. A beta particle consists of two protons and two neutrons.

_____ 5. The fission of U-235 can become a chain reaction because of all the energy liberated.

_____ 6. The process of uniting the nuclei of two light elements to form one heavier nucleus is known as nuclear fusion.

_____ 7. In radiocarbon dating, the ratio of C-14 to C-12 gives data relative to the age of the object being dated.

_____ 8. An atom of $^{235}_{92}U$ has 327 nucleons.

_____ 9. $^{232}_{92}U$ represents a uranium isotope with 92 protons and 146 neutrons.

_____ 10. The Curie is the unit of radiation that measures the rate of decay of radionuclides.

_____ 11. The half-life of P-30 is 3 minutes. Starting with a sample of the isotope that weighs 160 g, there would remain 32 g after a period of 15 minutes.

Multiple Choice (choose the best answer)

_____ 1. If $^{238}_{92}U$ loses an alpha particle, the resulting nuclide is

a) $^{237}_{92}U$ b) $^{234}_{90}Th$ c) $^{238}_{93}Np$ d) $^{210}_{83}Bi$

_____ 2. Which of the following is *not* a characteristic of nuclear fission?

a) Upon absorption of a proton, a heavy nucleus splits into two or more smaller nuclei.
b) Two or more neutrons are produced from the fission of each atom.

c) Large quantities of energy are produced.
d) All nuclei formed are radioactive, giving off beta and gamma radiation.

_____ 3. The half-life of Sn-121 is 10 days. If you started with 40 g of this isotope, how much would you have left 30 days later?

a) 10 g   b) None   c) 15 g   d) 5 g

_____ 4. The radioactivity ray with the greatest penetrating ability is

a) Alpha   b) Beta   c) Gamma   d) Proton

_____ 5. In a nuclear reaction

a) Mass is lost   b) Mass is gained
c) Mass is converted into energy   d) Energy is converted into mass

_____ 6. As the temperature of a radionuclide increases, its half-life

a) Increases   b) Decreases   c) Remains the same   d) Fluctuates

_____ 7. When $^{235}_{92}U$ is bombarded by a neutron, the atom can fission into

a) $^{124}_{53}I + ^{109}_{47}Ag + 2\,^{1}_{0}n$ b) $^{123}_{50}Sn + ^{110}_{42}Mo + 2\,^{1}_{0}n$

c) $^{134}_{56}Ba + ^{128}_{36}Xe + 2\,^{1}_{0}n$ d) No correct answer given.

_____ 8. In the nuclear equation $^{45}_{21}Sc + \,^{1}_{0}n \rightarrow X + \,^{1}_{1}H$, the nuclide X that is formed is

a) $^{45}_{22}Ti$ b) $^{45}_{20}Ca$ c) $^{46}_{22}Ti$ d) $^{45}_{20}K$

_____ 9. When $^{239}_{92}U$ decays to $^{239}_{93}Np$, what particle is emitted?

a) Positron   b) Neutron   c) Alpha particle   d) Beta particle

_____ 10. The rad is the unit of radiation that measures

a) An absorbed dose of radiation
b) Exposure to X-rays
c) The dose from a different type of radiation
d) The rate of decay of a radioactive substance

_____ 11. Which of the following is a correctly balanced equation for decay by beta emission?

a) $Al^{24}_{13} \rightarrow \beta + Mg^{24}_{12}$ b) $Al^{24}_{13} \rightarrow \alpha + Na^{20}_{11}$ c) $Al^{24}_{13} \rightarrow \beta + Na^{24}_{11}$ d) $Al^{24}_{13} \rightarrow \beta + S^{24}_{14}$

_____ 12. Which of these is *not* conserved in a balanced nuclear equation?

a) mass   b) atoms   c) energy   d) electrical charge

_____ **13.** Suppose that a particular nuclide absorbs two protons in a nuclear bombardment reaction, which of these statements is true?

    a) The chemical identity of the element will change.
    b) The mass of the element will decrease.
    c) The nuclear charge of the element will remain the same.
    d) The element's ionic charge will be +2.

Reasoning and Expression (use a separate sheet of paper)

1. Complete the following equations:

    a) $^{226}Ra \rightarrow {}^{222}Rn +$ _____   b) $^{10}B + n \rightarrow$ _____ $+ \alpha$
    c) $^{14}N + n \rightarrow$ _____ $+ H$ d) $^{56}Fe +$ _____ $\rightarrow {}^{57}CO + e^-$

2. Explain how a nucleus will be different if

    a) it emits an alpha particle
    b) it captures two neutrons
    c) it alpha decays twice and beta decays once

3. A sample of a particular isotope starts with a mass of 32 g. Twelve hours later, 0.5 g remains. What must be the half-life of this substance?

4. Consider the isotope: $^{90}_{38}Sr$

    a) How many neutrons does this isotope contain?
    b) What is the ratio of neutrons to protons?
    c) What will it become if it emits two protons?
    d) Explain briefly why it seems sensible, keeping in mind its placement on the periodic table, that this strontium isotope collects in bone tissue.

5. Element Z is known to have a mass number 102 and atomic number 45. When it radioactively decays, it emits first two alpha particles, then a beta, then another alpha, and finally another beta. What will be its charge and mass upon completion of this process?

6. Compare and contrast the relative mass and charge of

    a) an alpha particle and a proton
    b) a proton and a neutron
    c) a proton and an electron
    d) an alpha particle and a deuteron

7. Why is it so much easier for an atom to alter its electrical charge than its nuclear charge?

Name ______________________________
Section __________ Date ______________
Instructor ___________________________

CHAPTER 19 TEST

True — False

_____ **1.** Hydrocarbons are compounds composed entirely of carbon and hydrogen atoms bonded to each other with covalent bonds.

_____ **2.** Alkanes, alkenes, and alkynes are all hydrocarbons.

_____ **3.** Alkenes and alkynes are isomers.

_____ **4.** C_2H_2 and C_4H_8 belong to the same class of compounds.

_____ **5.** The IUPAC name for

$$\begin{array}{c}CH_3{-}\underset{\displaystyle CH_3}{\underset{|}{CH}}{-}\underset{\displaystyle CH_3}{\underset{|}{CH}}{-}CH_2{-}CH_3\end{array}$$

is 3,4-dimethylpentane.

_____ **6.** If Cl_2 is added to $CH_3CH = CH_2$, the product is $CH_3CH_2CH_2Cl$.

_____ **7.** When each member of a series of compounds differs from the next higher member by a CH_2 group, the series is called a homologous series.

_____ **8.** The IUPAC name for $CH_3CHClCH_3$ is chloropropane.

_____ **9.** Formaldehyde is $H{-}\overset{\displaystyle O}{\overset{\|}{C}}{-}H$.

_____ **10.** 2-Methylpentane and 2,3-dimethylbutane are isomers.

_____ **11.** The functional group in an organic compound may be polar or nonpolar, and it may contain a double or triple bond.

***Multiple Choice* (choose the best answer)**

_____ **1.** Which of the following is *not* a correct name for the alkane shown with it?

a) C_2H_6, ethane b) C_5H_{12}, propane c) C_7H_{16}, heptane d) $C_{10}H_{22}$, decane

_____ **2.** The product of the reaction $CH_3CH=CH_2 + H_2O + H^+ \rightarrow$ is an (a)

a) alcohol   b) aldehyde   c) alkyne   d) carboxylic acid

_____ **3.** The correct name for

$$CH_3\!-\!CH\!-\!CH\!-\!CH\!-\!CH=CH_2$$
with CH_3 and CH_3 substituents, and $CH_3\!-\!CH\!-\!CH_3$

a) Isobutane
b) 2,4-Methyl-3-propyl-5-hexene
c) 3,5-Dimethyl-4-isopropyl-l-hexene
d) 4,4-Diisopropylhexene

_____ **4.** Which of the following acids is named incorrectly?

a) $CH_3CH_2CH_2COOH$, butyric acid
b) $HCOOH$, formic acid
c) CH_3CH_2COOH, propic acid
d) CH_3COOH, acetic acid

_____ **5.** The number of isomers of C_6H_{14} is

a) 3   b) 5   c) 6   d) 8

_____ **6.** An open-chain hydrocarbon of formula C_6H_8 can have in its formula

a) one carbon-carbon double bond
b) two carbon-carbon double bonds
c) one carbon-carbon triple bond
d) one carbon-carbon double bond and one carbon-carbon triple bond

_____ **7.** The reaction $CH_2 = CH_2 + Br_2 \rightarrow CH_2BrCH_2Br$ represents

a) dehalogenation   b) substitution   c) addition   d) dehydration

_____ **8.** The general formula for a ketone is

a) RCHO   b) ROR   c) RCOOR   d) R_2CO

_____ **9.** Which of the following *cannot* be an aromatic compound?

a) C_6H_5OH b) C_6H_6 c) C_6H_{14} d) $C_6H_5CH_3$

_____ **10.** Which of the following pairs are not isomers?

a) CH_3OCH_2Cl and CH_2ClCH_2OH
b) CH_3CH_2CHO and $CH_3OCH_2CH_3$
c) $CH_3OCH_2OCH_3$ and $CH_3CH(OH)CH_2OH$
d) $C_6H_4(CH_3)_2$ and $C_6H_5CH_2CH_3$

_____ **11.** Which of these molecules can have no structural isomers?

      a) C_5H_{10} b) C_4H_{10} c) C_3H_8 d) C_6H_{12}

Reasoning and Expression (**use a separate sheet of paper**)

1. Write structural formulas for:

    a) 2,2-dichloro-3-methylbutane
    b) 2,3,3-trimethylhexane
    c) 1,2,2-trichloropropane

2. Complete these reactions:

        H H
        | |

    a) 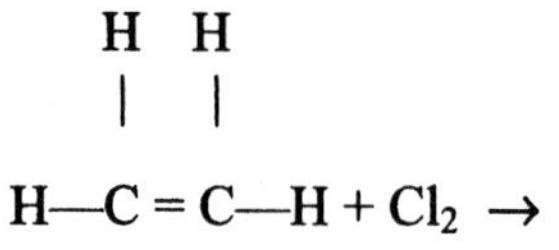 H—C = C—H + Cl_2 $\rightarrow$ b) $CH_3OH + HCOOH \xrightarrow{\;H^+\;}$

3. What is it about carbon that accounts for the large number of organic compounds found in nature?

4. Briefly explain the structural difference between:

    a) alkene and alkyne   b) alkane and alcohol   c) aldehyde and carboxylic acid

5. Sketch and name four isomers of $C_5H_{12}O$.

6. Someone has just announced the existence of a new molecule whose structure is suggested to be

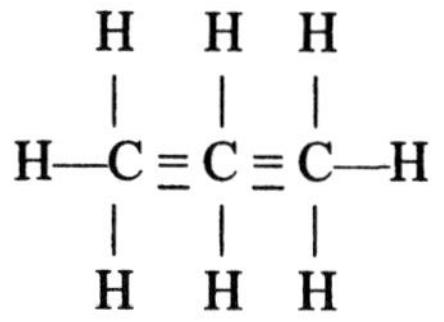

        H   H   H
        | | |
    H—C $\equiv$ C $\equiv$ C—H
        | | |
        H   H   H

Why do you have your doubts about this structure?

CHAPTER 20 TEST

True — False

_____ 1. Carbohydrates are polyhydroxy alcohols or polyhydroxy ketones, or compounds that will yield them when hydrolyzed.

_____ 2. The structure of a disaccharide molecule consists of two monosaccharide units linked together, minus a water molecule.

_____ 3. Starch and cellulose are both polysaccharides, but cellulose consists only of glucose units whereas starch is alternating glucose and fructose units.

_____ 4. Fats are digested in the stomach and converted to glucose.

_____ 5. Vegetable oils are hydrogenated by reacting them with sodium hydroxide.

_____ 6. Proteins are high molar-mass polymers of nucleotides.

_____ 7. The bond connecting the amino acids in a protein is commonly called a peptide linkage.

_____ 8. The flow of genetic information is in one direction, from DNA to RNA to proteins.

_____ 9. Enzymes are protein molecules that act as catalysts by greatly lowering the activation energy of specific biochemical reactions.

_____ 10. RNA is transcribed from DNA and controls the synthesis of proteins.

_____ 11. Saturated describes a fatty acid with either double or triple C-C bonds in its structure.

Multiple Choice (choose the best answer)

_____ 1. Sugars are members of a group of compounds with the general name

   a) Carbohydrates   b) Lipids   c) Proteins   d) Steroids

_____ 2. The products formed when maltose is hydrolyzed are

   a) Glucose and fructose   b) Glucose and galactose
   c) Glucose and glucose   d) Galactose and fructose

_____ **3.** Which is *not* true about glucose?

a) It is a monosaccharide.
b) It is a component of sucrose, maltose, lactose, starch, glycogen, and cellulose.
c) It is a ketohexose.
d) It is the main source of energy for the body.

_____ **4.** The sweetest of the common sugars is

a) Fructose   b) Sucrose   c) Glucose   d) Maltose

_____ **5.** Which of the following is *not* true?

a) Lactose is sweeter than sucrose.
b) Glycogen is known as animal starch.
c) Cellulose is the most abundant organic substance in nature.
d) Lactose is a disaccharide known as milk sugar.

_____ **6.** Which of the following lipids does not contain a glycerol unit as part of its structure?

a) A fat   b) A phospholipid   c) A glycolipid   d) An oil

_____ **7.** Which of the following amino acids contains sulfur?

a) Alanine   b) Histidine   c) Cysteine   d) Glycine

_____ **8.** Which of the following is *not* a correct statement about DNA and RNA?

a) DNA contains deoxyribose, whereas RNA contains ribose.
b) Both DNA and RNA are polymers made up of nucleotides.
c) DNA directs the synthesis of proteins and RNA contains the genetic code of life.
d) DNA exists as a double helix, whereas RNA exists as a single helix.

_____ **9.** Complementary base pairs in DNA are linked through the formation of

a) Phosphate ester bonds   b) Peptide linkages
c) Hydrogen bonds   d) Ionic bonds

_____ **10.** The process during which a cell splits to form a sperm or an egg cell is called

a) Mitosis   b) Meiosis   c) Translation   d) Transcription

_____ **11.** Which of these elements is found in proteins, but not in carbohydrates or fats?

a) oxygen   b) hydrogen   c) nitrogen   d) carbon

_____ **12.** The C-N connection in amino acid molecules is

a) an ionic bond   b) a nonpolar covalent connection
c) a peptide linkage   d) a triple bond

Reasoning and Expression (**use a separate sheet of paper**)

1. Explain briefly what each of these processes accomplishes.

    a) hydrolysis of disaccharides   b) saturation of an oil or fat   c) saponification

2. How do the species in the following pairs differ from one another?
    a) fats from fatty acids
    b) simple sugars from complex sugars
    c) saturated fats from unsaturated fats

3. How do the processes of photosynthesis and respiration differ from one another?
    How are they similar?

4. Suppose that the concentration of glucose in human red blood is about 0.0050 M.
    How many molecules of glucose are in a human body if it contains 5.8 liters of blood?

5. Compare and contrast proteins, peptides, and polypeptides.

Test Questions

The following questions arc compiled for the convenience of the instructor. The questions are of the objective type: i.e., true-false, multiple choice, matching, and completion. The questions are arranged by chapter.

These test questions assume the student has a periodic table available.

CHAPTER 1.   Introduction

True — False (choose one)

1.   Chemistry is a science that deals with the composition of matter and the changes in the composition of matter.

2.   Chemistry is an experimental science.

3.   The branch of chemistry concerned with the compounds containing the element carbon is inorganic chemistry.

4.   Substances classified as inorganic are derived mainly from mineral sources rather than from animal or vegetable sources.

5.   Chemistry and physics are overlapping sciences, since both are based on the properties and behavior of matter.

6.   Sound theoretical reasoning by the alchemists paved the way for the systematic experimentation that is characteristic of modern science.

7.   Scientific laws cannot be proven.

8.   A theory can never be proven correct, but one experiment can prove it incorrect.

9.   Hypothesis and theory are mental pictures to match observations and experimental results.

10.   A theory is the same as a hypothesis.

11.   A scientific law is a statement of fact about natural phenomena to which no exceptions are known under given conditions.

12.   William Buehler, in his development of "memory metal," observed that warm and cold metals differed in the sounds they emitted when dropped onto a surface.

13.   The parent shape of the memory metal nitinol is fixed by heating the material.

14.   Scientific breakthrough discoveries are rarely, if ever, the result of trial and error.

15.   Risk assessment requires a knowledge of both the severity of one's exposure to a chemical and the probability that exposure will occur.

16.   A realistic goal in risk management is to eliminate all risks of exposure to chemicals.

17. Advances in scientific research, such as Buehler's, illustrate that scientists often work alone, unassisted by others' experimental contributions.

18. In following the scientific method, it is important to test and retest one's hypotheses.

19. An experiment is a collection of fact and data performed under uncontrolled conditions.

20. Scientific laws are simple statements of natural phenomena to which no exceptions are known.

21. From 1803 to 1810, John Dalton advanced his atomic theory.

22. One of the principal goals of the alchemist was to change metals such as iron into gold.

23. Oxygen was discovered by Robert Boyle in 1774.

24. The use of the chemical balance revolutionized quantitative measurements in chemical reactions.

25. The two main branches of chemistry are organic and inorganic chemistry.

26. Only through well-planned expectation can great scientific discoveries be derived.

27. One may test a hypothesis by making measurements, noting color changes, or weighing the amount of a substance that forms in a reaction.

28. "Albert Einstein is the greatest scientist of the 20th century" is a testable hypothesis.

29. Communication between scientists is not important in the scientific method.

Multiple Choice (**choose the best answer**)

1. The statement "One apple a day keeps the doctor away" is best described as a(n)

    a) hypothesis   b) observation   c) theory   d) myth

2. The statement "Mass remains constant during chemical reactions" is best described as a(n)

    a) observation   b) natural law   c) theory   d) hypothesis

3. The statement "Nitrol is an intelligent material" is best described as a(n)

    a) theory   b) observation   c) hypothesis   d) experiment

4. All of the following apply to hypothesis, theory, and law except

    a) A theory is a tested explanation of natural phenomena.
    b) Useful hypothesis leads to new experiments.
    c) A law summarizes a theory.
    d) Laws are simple statements of natural phenomena.

5. The statement "An atom consists of a dense nucleus surrounded by clouds of electrons" is best described as a(n)

a) observation   b) natural law   c) theory   d) hypothesis

CHAPTER 2.   Standards of Measurement

True — False (choose one)

1. The terms "mass" and "weight" are synonymous and both are a measure of the earth's gravitational attraction for a body.

2. The mass of an object decreases as the distance between the object and the center of the earth increases.

3. The number of significant figures in the product 124.8 x 62.0 x 15 is two.

4. The number 2,486.0 expressed in scientific notation is 2.486×10^3.

5. The number 64.6052 rounded off to four digits is 64.60.

6. "Include units in every problem you solve" is the first law of dimensional analysis.

7. Reporting the number in a measurement is more important than reporting the unit.

8. The number of significant figures in 1.001 is the same as the number of significant figures in 1.100.

9. The metric system is a decimal system of units for measurements of mass, length, time, and other physical constants.

10. The meter and the kilogram are two of the base units of the SI system of measurement.

11. The standard unit of mass in the metric system is the kiloliter.

12. The number of centimeters in one foot is more than the number of inches in one yard.

13. A nanometer is smaller than a micrometer.

14. 1255 mg is equal to 1.255 g.

15. One mL equals 1 cm^3 exactly.

16. Heat and temperature are forms of energy.

17. A Fahrenheit degree is larger than a Celsius degree.

18. 70°F is hotter than 24°C.

19. 32°C is equal to 0°F.

20. 100°C is equal to 212°F.

21. Water and air are common reference materials for comparing densities.

22. The volume of 20.0 g of gold is greater than the volume of 10.0 g of lead. (density Au = 19.3 g/mL; density Pb = 11.3 g/mL).

23. The unit for the answer to the calculation, 2.4 cm x 2.4 cm x 2.4 cm is cm.

24. The density of gases is normally expressed in terms of grams per cubic centimeter (g/cm^3) or grams per milliliter (g/mL).

25. A decrease in temperature often causes a substance to shrink and increase its density.

26. An increase in temperature often causes a material to expand and increase its mass.

27. The unit of measurement cannot be treated as algebraic quantities in calculation.

28. If a 5.0 g object has a volume of 10.0 ml then its density is 2.0 g/mL.

29. If ice floats in water and sinks in ethyl alcohol then the density of ethyl alcohol is less than the density of water.

30. One liter of ethyl alcohol, density 0.79 g/mL, has a greater mass than one quart of water, d = 1.0 g/mL.

31. The density of water in the solid state is denser than liquid water.

32. A graduated cylinder, a buret, and a pipet are instruments for measuring the volume of liquids.

33. A degree Celsius is a smaller unit than a Kelvin.

34. The measured temperature of a material does not depend upon how much of it you have to work with.

35. The SI unit liter most closely resembles the English unit gallon in volume.

36. A milligram is 0.001 g.

37. A centimeter is longer than a millimeter.

38. If 1 mL = 0.001 liter, then we can use the factor 10^3 mL/liter to convert liters to milliliters.

39. The density of water at 4°C is 1.00 g/mL.

40. The density of an object is fixed and independent of its temperature.

41. The measurement 12.200 g contains three significant figures.

42. The answer to 25.2 x 0.1465 should contain three significant figures.

43. The number 14.0667 rounded off to four digits is 14.07.

44. The answer to 16.215 - 2.32 should contain three digits.

45. A liter contains 100 mL.

46. 90°C is hotter than 210°F.

47. The units of specific gravity are g/mL.

48. The mass of an object is fixed and independent of its location.

49. The prefix *milli* means one-hundredth of.

50. The number 0.002040 written in scientific notation is 2.04×10^{-3}.

51. The density of liquid A is 2.20 g/mL, and that of liquid B is 1.44 g/mL. When equal volumes of these two immiscible liquids are mixed, liquid A will float on liquid B.

52. The prefixes milli, centi, deka, and kilo are listed in order of increasing magnitude.

53. The density of a milliliter of ocean water is less than the density of a kiloliter of the same water.

54. The temperature of a milliliter of ocean water is less than the temperature of a kiloliter of the same water.

55. The heat contained in a milliliter of ocean water is less than the heat of a kiloliter of the same water.

Multiple Choice **(choose the best answer)**

1. The number of significant figures in 0.001040 is

    a) 7   b) 6   c) 4   d) 3

2. When multiplying 1.04 x 0.31 the number of significant figures in the answer should be

    a) 2   b) 3   c) 4   d) 5

3. How many significant figures should be in the answer obtained by adding 0.04, 31.2, and 100.0?

    a) 3   b) 2   c) 5   d) 4

4. $\dfrac{10^1}{10^3} = ?$

    a) 10^4 b) 10^2 c) 10^{-2} d) 10^{-4}

5. $10^n \times 10^m = ?$

    a) $10^{(n-m)}$ b) $10^{(m-n)}$ b) 10^{nm} d) $10^{(n+m)}$

6. Which prefix is followed by an incorrect numerical equivalent?

    a) kilo, 10^3 b) deci, 10^1 c) Milli, 10^{-3} d) Centi, 10^{-2}

7. The prefix used for 10^{-6} is

    a) milli   b) micro   c) nano   d) mega

8.   Which unit is not a basic unit of measurement in the metric system (International System)?

      a) gram   b) Kelvin   c) meter   d) kilogram

9.   Which unit is not a basic unit of measurement in the International System?

      a) meter (length)   b) second (time)   c) °C (temperature)   d) mole (amount of substance)

10.   The SI unit for mass is

      a) gram   b) kilogram   c) milligram   d) none of these

11.   When multiplying the measurements 4.235 x 12.60 x 0.143 the answer, using the correct number of significant figures, is

      a) 7.631   b) 7.6   c) 7.63   d) 7.6306

12.   Of the following which one represents the greatest speed?

      a) 55 mi/hr   b) 55 km/hr   c) 55 ft/sec   d) 55 m/sec

13.   In the metric system the prefix for one hundred is

      a) hecto   b) centi   c) deca   d) mega

14.   The number of centimeters in 4.30 ft is

      a) 10.9 cm   b) 30.5 cm   c) 131 cm   d) 151 cm

15.   The number of inches in one centimeter is

      a) 2.54   b) 0.394   c) 12.0   d) 0.100

16.   Which expression is incorrect?

      a) 4.54 lb = 1 kg   b) 2.54 cm = 1 inch   c) 1,000,000 mg = 1kg   d) 1.00 mL = 1.00 cm^3

17.   One centimeter is equal to

      a) 2.54 in.   b) 0.328 ft   c) 0.109 yd   d) No correct answer given.

18.   The number of miles in one kilometer is

      a) 1000   b) 0.621   c) 1.61   d) 6.21

19.   The number of milligrams in 10.0 g is

      a) 100   b) 1.00 x 10^{-3} c) 1.00 x 10^4 d) 1.00 x 10^3

20. What is the cost of one kilogram of sugar if sugar costs $1.37 per 5.0 lbs?

      a) $0.27   b) $0.60   c) $3.11   d) $2.74

21. How many aspirin tablets can be made from 100 g of aspirin if each tablet contains 5.0 grains of aspirin? (7000 grains are equal to one pound.)

      a) 80   b) 320   c) 32   d) 308

22. Convert 2.4 lb/in^3 to kg/m^3.

      a) 6.7×10^7 kg/m^3 b) 4.3×10^1 kg/m^3 c) 6.7×10^4 kg/m^3 d) 4.3×10^4 kg/m^3

23. How many minutes would it take a car traveling at 55 miles/hour to cover 20.0 km? (1 mile =1.609kn)

      a) 25.0 min   b) 21.8 min   c) 13.56 min   d) 18.3 min

24. The number of grams in 12.5 mg is

      a) 12.5×10^3 b) 1.25×10^2 c) 0.125   d) 1.25×10^{-2}

25. The conversion factor to change milligrams to grams is

      a) $\dfrac{1000\ \text{mg}}{1\text{g}}$ b) $\dfrac{100\ \text{mg}}{1\text{g}}$ c) $\dfrac{1\text{g}}{100\ \text{mg}}$ d) $\dfrac{1\text{g}}{1000\ \text{mg}}$

26. How many pounds does 5.00 gallons of water weigh?

      a) 8.34 lb   b) 17.2 lb   c) 41.7 lb   d) No correct answer given.

27. The volume of 1 liter is

      a) equal to 1 quart   b) less than 1 quart   c) greater than 1 quart   d) 0.25 gallon

28. One liter is equal in volume to

      a) 1 quart   b) 946 mL   c) 1000 mL   d) 0.01 kL

29. Which formula can be used to convert °C to K?

      a) °C = K + 273   b) °C = 1.8°F + 32   c) K = $\dfrac{°F - 32}{1.8}$ + 273   d) No correct answer given.

30. Which temperature change is the smallest?

      a) 0K → 10K   b) 0°F → 10°F   c) 0°C → 10°C   d) 0°F → 10°F

31. 255°F is equivalent to

    a) 114°C   b) 397 K   c) 528 K   d) No correct answer is given.

32. 120°C is equivalent to

    a) 248°F   b) 216°F   c) 489 K   d) No correct answer is given.

33. Two objects, A and B, have identical volumes. The mass of A is 17.0 g and the mass of B is 14.0 g. Which statement is correct?

    a) the density of B is greater than the density of A
    b) the density of A is the same as the density of B
    c) Density B = Density A x $\frac{14.0}{17.0}$
    d) Density B = Density A x $\frac{17.0}{14.0}$

34. The distance between atoms is often expressed in Angstrom (Å) (1 Å = 1 x 10^{-10} m). The distance between the two hydrogen atoms in H_2 is 7.4 x 10^{-9} cm. Express this distance in Å.

    a) 7.4 x 10^{-1} Å   b) 7.4 X 10^{-1} Å   c) 7.4 x 10^{-2} Å   d) 7.5 x 10^{-2} Å

35. A car gets 35.0 miles per gallon. How far can it travel on 30.0 liters of gasoline? (1 gallon = 3.785 liters)

    a) 277 miles   b) 362 miles   c) 145 miles   d) 727 miles

36. 20 g of liquid A are mixed with 50 g of liquid B. The density of the resulting solution is 1.2 g/ml. Which statement must be correct?

    a) The density of liquid A is greater than the density of liquid B.
    b) The density of liquid B is greater than the density of water.
    c) Both of the liquids are denser than water.
    d) At least one of the liquids is denser than water.

37. The density of copper is 8.92 g/ml. The mass of a piece of copper that has a volume of 10.0 mL is

    a) 0.892 g   b) 892 g   c) 89.2 g   d) 8.9 x 10^2 g

38. The volume occupied by 20.0 g of (density 13.6 g/ml) is

    a) (20.0 x 13.6)mL   b) $\frac{20.0}{13.6}$ mL   c) $\frac{13.6}{20.0}$ mL   d) (20.0 - 13.6) mL

39. An empty graduated cylinder weighs 54.772 g. When filled with 50.0 mL of a liquid it weighs 97.701 g. The density of the liquid is

    a) 1.95 g/mL   b) 1.10 g/mL   c) 0.910 g/ml   d) 0.859 g/ml

40. Platinum has a density of 21.45 g/cm^3. A cube of platinum, 4.00 cm on each side, has a mass of

    a) 579 g   b) 1.37 x 10^3 g   c) 343 g   d) 686 g

41. The rule for recording significant figures is to use all digits which are known or certain plus those digits which are estimated.

      a) 1   b) 2   c) 3   d) 4

42. The number 14.8946 rounded to four digits becomes

      a) 14.80   b) 14.90   c) 14.89   d) 14.88

43. The number 6.996 rounded to three digits becomes

      a) 6.99   b) 7.00   c) 6.90   d) 6.00

44. 1.00 cm is equal to how many inches?

      a) 0.394   b) 0.10   c) 12   d) 2.54

45. 42.0°C is equivalent to

      a) 273 K   b) 5.55°F   c) 108°C   d) 53.3°F

46. An object has a mass of 62 g and a volume of 4.6 mL. Its density is

      a) 0.074 mL/g   b) 285 g/mL   c) 7.4 g/mL   d) 13 g/mL

47. The conversion factor to change grams to milligrams is

      a) $\dfrac{100\ mg}{1g}$ b) $\dfrac{1g}{100\ mg}$ c) $\dfrac{1g}{1000\ mg}$ d) $\dfrac{1000\ mg}{1g}$

48. What Fahrenheit temperature is twice the Celsius temperature?

      a) 64°F   b) 320°F   c) 200°F   d) 746°F

49. A gold alloy has a density of 12.41 g/ml and contains 75.0% gold by mass. The volume of this alloy that can be made from 255 g of pure gold is

      a) 4.22×10^3 mL   b) 2.37×10^3 mL   c) 27.4 mL   d) 15.4 mL

50. A 22.4 g piece of aluminum is dropped in a graduated cylinder containing 24.0 mL of water. The water level rises to 32.3 mL. The density of aluminum is

      a) 2.70 g/mL   b) 0.371 g/mL   c) 5.62 g/mL   d) 1.84 g/mL

51. Perform the following operation to the correct number of significant figures.

$$\frac{75.032+2.3+0.0046}{10.2}$$

      a) 7.6   b) 7.57   c) 7.58   d) 7.578

52. A lead cylinder ($V = \pi r^2 h$) 12.0-cm in radius and 44.0-cm long has a density of 11.4 g/mL. The mass of the cylinder is

   a) 2.27×10^5 g   b) 1.89×10^5 g   c) 1.78×10^3 g   d) 3.50×10^5 g

53. The recommended adult dosage of a particular medication used to treat asthma is 6.0 mg per kg of body weight. What dose, in milligrams, is required for a 170-pound person?

   a) 0.035 mg   b) 460 mg   c) 1020 mg   d) 2250 mg

54. Expressed to the correct number of significant digits, the answer to the calculation

   $$\frac{(5.031 - 4.96)(2.38)}{391}$$ is

   a) 0.04   b) 0.043   c) 0.0432   d) 0.04322

55. Which measurement contains three significant figures?

   a) 3.00 g/mL   b) 0.30 g/mL   c) 0.03 g/mL   d) 0.003 g/mL

56. A piece of silver foil has a thickness of 0.0119 cm and a width of 1.00 cm. If the density of silver is 10.5 g/cm^3, how long should the piece be to give exactly 3.00 g of the silver?

   a) 0.286 cm   b) 3.57 cm   c) 24.0 cm   d) 37.5 cm

57. A rectangular chunk of an unknown alloy measures 8.0 cm by 1.0 m by 14.0 mm. Its mass is 2.0 kg. Its density is correctly expressed as

   a) 2.0 g/cm^3 b) 1.8 g/cm^3 c) 1.79 g/cm^3 d) 1.785 g/cm^3

58. All of the following statements are correct regarding weight and mass except:

   a) Weight is a measure of the earth's gravitational pull on a substance.
   b) Mass is the quantity of matter in a substance.
   d) Mass and weight have the same value in the absence of gravity.
   c) The weight of an object changes with altitude.

Matching

From the list given below, place the letter of the appropriate term in front of each definition.

List of terms: (a) temperature; (b) kilogram; (c) meter; (d) density; (e) mass; (f) weight; (g) balance;
(h) specific gravity; (i) thermometer; (j) significant figures; (k) gram; (l) graduated cylinder

_____ **1.** An instrument used to measure the temperature of a system.

_____ **2.** The quantity of matter that a body possesses.

_____ **3.** The unit of mass in the SI system.

_____ **4.** The mass per unit volume of a substance.

_____ **5.** The measure of the gravitational attraction between the earth and an object.

_____ **6.** An instrument used to measure the mass of an object.

_____ **7.** The digits expressing the known precision of a measured quantity.

_____ **8.** A measure of the intensity of heat, or how hot a system is.

_____ **9.** The ratio of the density of a substance to the density of another substance.

_____ **10.** The standard metric unit of length.

CHAPTER 3. Classification of Matter

True — False (choose one)

1. Matter is anything that has mass and occupies space.

2. Solids with definite crystalline structures are said to be amorphous.

3. Plastics, glass, and gels are examples of crystalline solids.

4. A substance is homogeneous and has a fixed composition.

5. A solution of sodium chloride in water is an example of a substance.

6. A homogeneous mixture will consist of two or more phases.

7. A substance may be a homogeneous mixture.

8. A liquid has definite volume but no definite shape.

9. A mixture of ice and water is homogeneous.

10. The two forms of pure substances are elements and compounds.

11. Of the three states of matter, a gas has its particles farthest apart.

12. A gas has a definite volume but no definite shape.

13. Ice is a different substance than water because it is solid while water is liquid.

14. Amorphous solids do not have crystalline structures.

15. A solution is a homogeneous mixture.

16. Elements are the basic building blocks of all substances.

17. No elements other than those on earth have been detected on other bodies in the universe.

18. The smallest particle of an element that can exist is the molecule.

19. An atom is the smallest unit of an element that can enter into a chemical reaction.

20. Elements are distributed equally in nature.

21. The most abundant element in the earth's crust is also the most abundant element in the human body.

22. A knowledge of symbols is essential for writing chemical formulas and equations.

23. The correct pronunciation for the formula NaCl is "nacle."

24. Compounds can be decomposed chemically into simpler substances.

25. The symbol for iron is Fe.

26. Compounds consisting of ions do not exist as molecules.

27. Atoms are held together in compounds by chemical bonds.

28. The nine million known chemical compounds represent most of the compounds that can possibly exist.

29. The symbols for most of the elements contain one or two letters.

30. The symbol for potassium is P.

31. The symbol for calcium is Ca.

32. The symbols for sodium, silver, and lead are Na, Si, and Pb, respectively.

33. There are more nonmetal elements than metal elements.

34. Many metals readily combine chemically with one another.

35. Nonmetals frequently combine with one another to form ionic compounds.

36. All diatomic molecules contain two atoms.

37. All the elements that exist as diatomic molecules are gases at room temperature.

38. Two of the elements, bromine and mercury, are liquids at room temperature.

39. A chemical formula shows the symbols and the ratio of the atoms of the elements in a compound.

40. The formula $Ca_3(PO_4)_2$ indicates that the compound contains six atoms of oxygen.

41. Both benzene, C_6H_6, and acetylene, C_2H_2, have the same ratio of carbon to hydrogen in their molecules.

42. A compound is a pure substance and cannot be separated by physical means.

43. A substance is homogeneous but does not have a fixed composition.

44. A system having more than one phase is heterogeneous.

45. Plastics, glass, and gels are examples of amorphous solids.

46. The most abundant element in the earth's crust, seawater, and atmosphere is nitrogen.

47. The symbol for cobalt can be written as Co or CO.

48. Metalloids are elements that have properties intermediate between those of metals and nonmetals.

49.   Compounds exist as either molecules or ions.

50.   A mixture is a combination of two or more substances in which the substances retain their identity.

51.   Pure substances occur in two forms, elements and ions.

52.   Elements that have properties resembling both metals and nonmetals are called mixtures.

53.   The symbols of the elements have two letters.

54.   In general, solids are denser than liquids, and liquids are denser than gases.

55.   The basic building blocks of all substances, which cannot be decomposed into simpler substances by ordinary chemical change, are compounds.

56.   The smallest particle of an element that can exist and still retain the properties of the element is called an atom.

57.   The symbol for silver is Ag.

58.   The symbol for nitrogen is Ni.

59.   The smallest uncharged unit of a compound is an atom.

60.   All elements are expressed as diatomic molecules in the gaseous state.

61.   A positively charged atom or group of atoms is called an *anion.*

62.   All ions carry an electrical charge.

63.   The symbols Na, Ni, and Ne all represent metallic elements.

64.   Liquid substances always possess a definite shape.

65.   Fluorine is the most reactive nonmetal and combines readily with most all other elements.

66.   Gases have density intermediate between liquids and solids.

67.   All homogeneous samples are liquid solutions.

68.   The dollar coin is a homogeneous solution.

Multiple Choice (choose the best answer)

1.   Which of the following is *not* true of a gas?

    a) A gas is able to be compressed into a very small volume.
    b) The particles of a gas are held together by very strong attractive forces but not in a rigid form.
    c) A gas presses continuously and in all directions on the vessel walls.
    d) Gas particles fill the entire space in which they are contained.

2.   A substance which always fills its container is

a) gas   b) a liquid   c) a solid   d) amorphous matter

3.   Air is classified as a mixture because it

a) may be liquefied   b) does not have a definite composition
c) has a density of 1.29 g/mL   d) is colorless and odorless

4.   Which statement is *not* true about a system containing steam and water?

a) the system is homogeneous   b) the system is heterogeneous
c) the system contains two phases   d) the system contains one kind of matter

5.   Which element is among the most abundant in the human body but not among the five most abundant elements in the earth's crust, seawater, and atmosphere?

a) phosphorous   b) carbon   c) aluminum   d) oxygen

6.   So is the symbol of

a) sodium   b) solenium   c) osmium   d) none of the elements

7.   Co is the symbol for

a) copper   b) cobalt   c) carbon monoxide   d) carbon

8.   Which of the following represents an element?

a) CO   b) NI   c) N_2 d) OS

9.   Which of the following is *not* a characteristic of metals?

a) always a solid at room temperature   b) malleable
c) good conductor of heat   d) good conductor of electricity

10.   Which of the following is *not* a characteristic of nonmetals?

a) reacts with metals   b) reacts with other nonmetals
c) brittle when in solid state   d) shiny

11.   Which of the following is *not* a mixture?

a) air   b) salt water   c) milk   d) sugar

12.   Which of the following is a compound?

a) lead   b) water   c) potassium   d) tea

13. Which of the following elements occur as diatomic molecules?
I. Helium   II. Hydrogen   III. Sulfur   IV. Oxygen

    a) II, IV   b) I, II, IV   c) II only   d) All of the above.

14. Which two expressions in each of the following sets represent the same number of oxygen atoms?

    a) H_2O and $2 H_2O$ b) $2 H_2O_2$ and K_2CrO_4 c) $2 OH^-$ and H_2O

    d) $C_{12}H_{22}O_{11}$ and $2 C_6H_{12}O_6$

15. In the formula $(NH_4)_2C_2O_4$

    a) there are more nitrogen atoms than oxygen atoms
    b) there are 4 carbon atoms
    c) there are 8 oxygen atoms
    d) the number of carbon atoms is the same as the number of nitrogen atoms

16. Which of the following formulas contains the fewest oxygen atoms?

    a) $Ba(ClO_3)_2$ b) $K_2Cr_2O_7$ c) $Na_2CO_3 - 10 H_2O$ d) $Ca(MnO_4)_2$

17. The number of oxygen molecules in $Ba(OH)_2 - 8 H_2O$ is (be careful)

    a) 2   b) 0   c) 8   d) 10

18. The formula $C_3H_5(OH)_3$ contains

    a) 6 carbon atoms   b) 9 carbon atoms   c) 18 hydrogen atoms   d) a total of 14 atoms

19. The formula of the compound that contains two atoms of nitrogen and five atoms of oxygen per formula is

    a) $2 N_5O$ b) NO c) N_2O_5 d) N_5O_2

20. Which of the following is an element?

    a) iron   b) wood   c) water   d) blood

21. How many atoms of oxygen are represented by the formula $Fe(NO_3)_3$?

    a) 3   b) 6   c) 9   d) 12

22. The formula for hydrochloric acid is

    a) HCl b) $HClO$ c) $HClO_2$ d) H_2Cl

23. Which of the following is *not* one of the five most abundant elements by mass in the earth's crust, seawater, and atmosphere?

    a) Oxygen   b) Hydrogen   c) Silicon   d) Aluminum

24. Which of the following is a mixture?

    a) Water   b) Chromium   c) Wood   d) Sulfur

25. Which of the following could be a substance?
I. Compounds   II. Mixtures   III. Elements   IV. Solutions

    a) I, III, IV   b) I, III   c) I, V   d) All of the above.

26. A colleague gave you the formula of the following compound on the phone, CA-N-O3 taken twice. The correct written formula is

    a) $Ca(NO)_6$ b) $CaNO_6$ c) $Ca(NO_3)_2$ d) $2\,Ca(NO_3)$

27. A colleague gave you the formula of the following compound on the phone, C6-H12-O6. The correct written formula is

    a) $(CH_2)O_6$ b) $(CO)_6\,H_{12}$ c) $C_6(H_2O)_6$ d) $C_6H_{12}O_6$

28. When a pure substance was analyzed, it was found to contain carbon and chlorine. This substance must be classified as

    a) An element   b) A mixture   c) A compound   d) Both a mixture and a compound

29. Sodium, carbon, and sulfur have the symbols

    a) Na, C, S   b) So, C, Su   c) Na, Ca, Su   d) So, Ca, Su

30. The number of oxygen atoms in $Al(C_2H_3O_2)_3$ is

    a) 2   b) 3   c) 5   d) 6

31. Which of the following is a mixture?

    a) Water   b) Iron (II) oxide   c) Sugar solution   d) Iodine

32. Which is the most compact state of matter?

    a) Solid   b) Liquid   c) Gas   d) Amorphous

33. A chemical formula is a combination of

    a) Symbols   b) Atoms   c) Elements   d) Compounds

34. Two molecules of fructose, $C_6H_{12}O_6$, contain

    a) an equal number of all atoms   b) two atoms of fructose
    c) twelve atoms of carbon   d) 24 total atoms

35. The symbol for one crystalline form of sulfur is S_8. That symbol shows

    a) one molecule composed of eight atoms
    b) eight molecules
    c) one atom composed of eight molecules
    d) one ion containing eight electrical charges

36. Which formula shows a compound which comprises four nonmetallic elements?

    a) HNO_3 b) $(NH_4)_3PO_4$ c) $CaSO_4 - H_2O$ d) P_2O_3

37. Which one of the following statements is incorrect?

    a) Heterogeneous mixture can be separated by physical means.
    b) A solution always contains a liquid phase.
    c) Sterling silver is a homogeneous mixture.
    d) Sodium chloride (table salt) dissolved in water forms a homogeneous mixture.

38. Which one of the following statements is true?

    a) Every substance is a compound.
    b) All mixtures are homogeneous.
    c) Every mixture contains two or more free elements.
    d) Every mixture contains two or more substances.

39. Which one of the following statements is false?

    a) Every compound is a substance.
    b) Every mixture contains two or more compounds.
    c) Every compound contains two or more elements.
    d) Every substance is a compound.

40. An alloy is

    a) a physical property
    b) a homogeneous solution of two different metals
    c) a dollar bill
    d) a homogeneous solution of a metal and nonmetal

Matching

From the list of terms given, choose a term to correctly identify each phrase.

List of terms: (a) compound; (b) symbols; (c) elements; (d) molecule; (e) ion; (f) chemical formulas; (g) metalloid; (h) atom; (i) amorphous; (j) cation; (k) anion; (1) mixture; (m) heterogeneous; (n) homogeneous; (o) solid; (p) alloy

_____ 1.   The smallest uncharged individual unit of a compound formed by the union of two or more atoms.

_____ 2.   A positive or negative electrically charged atom or group of atoms.

_____ 3.   A substance that cannot be broken down, by chemical means, to simpler substances.

_____ 4.   The smallest particle of an element that can exist.

_____ 5.   Abbreviations of the names of elements.

_____ 6.   Abbreviations used to represent compounds.

_____ 7.   A substance containing two or more elements combined in a definite proportion by weight.

_____ 8.   An element whose properties are intermediate between the properties of metals and nonmetals.

_____ 9.   A positively charged ion.

_____ 10.   Without shape or form.

_____ 11.   Consisting of two or more physically distinct phases.

_____ 12.   A form of matter that has definite shape and volume.

_____ 13.   A homogeneous solution of two nonmetals.

CHAPTER 4. Properties of Matter

True — False (choose one)

1. All moving bodies possess kinetic energy.

2. An energy transformation occurs whenever there is a chemical change.

3. The transformation of energy from one form to another is contrary to the law of conservation of energy.

4. Kinetic energy is the energy that matter possesses due to its position.

5. When sodium and chlorine react to produce sodium chloride and heat, the product, sodium chloride, will be at a higher potential energy than the reactants.

6. The vaporization of liquid water is a chemical change.

7. Matter can have both potential and kinetic energy.

8. Whenever a chemical change occurs a physical change occurs also.

9. In a chemical change there is a change in the composition of matter.

10. A change in the state of matter is a physical change with an accompanying change in composition.

11. In a chemical change the mass of the reactants equals the mass of the products.

12. Digesting food is a chemical change and the rusting of iron is a physical change.

13. When copper reacts with oxygen to form copper(II) oxide, a chemical change occurs.

14. Color, taste, odor, density, and boiling point are all physical properties.

15. Chemical equations are shorthand methods of expressing chemical changes.

16. Substances are recognized and identified by their physical and chemical properties.

17. The SI unit for energy is the calorie.

18. A physical change is reversible.

19. A chemical change is always accompanied by a physical change.

20. A physical change is always accompanied by a chemical change.

21. A chemical equation is a molecular representation of a chemical and physical change.

22. The specific heat of a substance is the quantity of heat required to change the temperature of 1 g of the substance by 1°C.

23. One calorie is equal to 4.184 joules.

24. One joule is equal to 4.184 calories.

25. The nutritional or large Calorie is equal to 10 calories.

26. 1.00 g Cu (specific heat = 0.385 J/g°C) will absorb more heat energy per degree than 1.00 g Fe (specific heat = 0.473 J/g°C).

27. The specific heat is the amount of heat energy required to raise 100 g of water by 1°C.

28. 41.84 J will raise the temperature of 10.0 g H_2O by 1°C.

29. An unknown compound is heated; a dark purple vapor and white crystalline solid result. This is an example of a chemical change.

30. When a platinum wire is heated red hot, only a physical change occurs.

31. A calorie is a larger unit of heat than a joule.

32. All chemical changes either lose or take in energy.

33. A leaf falling from a tree (before it hits the ground) demonstrates the process of kinetic energy changing into potential energy.

34. A roller coaster car climbing up a steep ramp and slowing to a stop as it climbs illustrates kinetic energy being changed into potential energy.

35. In a chemical change, substances are formed that are entirely different, having different properties and composition from the original material.

36. The Law of Conservation of Energy says that, because of the energy shortage, anyone wasting energy can be arrested.

37. The starting substances in a chemical reaction are called the reactants.

38. When a clean copper wire is heated in a burner flame it gains mass.

39. The energy released when hydrogen and oxygen react to form water was stored in the hydrogen and oxygen as chemical or kinetic energy.

40. In a physical change the composition of matter does not change.

41. Physical properties describe the ability of a substance to form new substances.

42. All matter can undergo both physical and chemical changes.

43. The specific heat of a substance is the quantity of heat lost when the temperature of 1g of that substance drops 1.0°C.

44. If 840 g of an ore of iron and oxygen yields 200. g of Fe, the ore is 76% oxygen.

45. Both mass and energy are conserved in a chemical change.

46. Joules and calories are both common units to measure temperature.

47. In a chemical change, the mass of the products is less than the mass of the reactants.

48. Specific heat actually measures the ability of a substance to change temperature; the greater a substance's specific heat, the easier it is to warm and cool it.

Multiple Choice **(choose the best answer)**

1. Which of the following is *not* a physical property?

    a) boiling point   b) density   c) digesting food   d) color

2. Which of the following is a physical change?

    a) a piece of sulfur is burned   b) a nail rusts   c) ice melts   d) a potato rots

3. Which of the following is a chemical change?

    a) water evaporates   b) ice melts   c) rocks are ground to sand   d) silverware tarnishes

4. Which is a chemical property of chlorine?

    a) It is a yellowish-green gas   b) It boils at - 34.6°C
    c) It combines with sodium to form sodium chloride   d) It has a sharp, suffocating odor

5. An example of a chemical property is

    a) TNT is an explosive
    b) gasoline is flammable
    c) Zn dissolves in hydrochloric acid to produce hydrogen gas
    d) All of the above.

6. According to the Law of Conservation of Energy, energy

    a) can be created
    b) can be destroyed
    c) can be created and destroyed
    d) can be converted from one form to another

7. When 12.66 grams of calcium are heated in air, 17.73 g of calcium oxide is formed. The percent of oxygen in the compound is

    a) 28.6%   b) 40.4%   c) 71.4%   d) 1.40%

8. Mercury(II) sulfide, HgS, contains 86.2% mercury. The mass of HgS that can be made from 60.0 g of mercury is

   a) 3448 g   b) 0.464 g   c) 34.5 g   d) 69.6 g

9. Magnesium chloride is 25.5% magnesium. What mass of magnesium could be recovered by the decomposition of 25.5g of magnesium chloride?

   a) 25.5 g   b) 35.0 g   c) 6.50 g   d) 8.94 g

10. If 8.50 g of potassium combines with 7.71 g of chlorine to form potassium chloride, what is the percent potassium in the potassium chloride?

    a) 52.4%   b) 64.8%   c) 1.54%   d) 60.7%

11. The percentage of oxygen in $MgSO_4 \cdot 7 H_2O$ is

    a) 71.4%   b) 26.0%   c) 32.5%   d) 75.6%

12. Barium iodide, BaI_2, contains 35.1% barium. A 13.0 g sample of barium iodide contains how many grams of iodine?

    a) 8.44 g   b) 4.56 g   c) 35.1 g   d) 3.51 g

13. Zinc oxide contains 80.3% zinc. What size sample would contain 5.35 g of zinc?

    a) 6.66 g   b) 27.2 g   c) 4.30 g   d) 11.0 g

14. When 15.0 g Mg and 15.0 g S were mixed and reacted to give the compound magnesium sulfide (MgS), 3.5 g of Mg remained unreacted. What percentage of the compound is magnesium (Mg)?

    a) 50.0%   b) 56.8%   c) 43.2%   d) 31.6%

15. Specific heat of Fe = 0.473 J/g°C; specific heat of Pb = 0.0305 J/g°C. To raise the temperature of 1.0 g of these metals by 1°C,

    a) both metals require the same amount of heat energy
    b) Fe requires more energy than does Pb
    c) Pb requires more energy than does Fe
    d) No correct answer given.

16. What is the specific heat of a solid that requires 124 J to raise the temperature of 95.5 g of the solid from 20.0°C to 25.0°C?

    a) 2.7 J/g°C   b) 0.27 J/g°C   c) 1.4 J/g°C   d) 5.9×10^4 J/g°C

17. How many calories are required to raise 125 g of water from 24.0°C to 42.5°C?

    a) 1.44×10^2 cal   b) 1.25×10^2 cal   c) 9.68×10^3 cal   d) 2.31×10^3 cal

18. How many joules of energy are absorbed when 250. g Pb (specific heat = 0.0305 J/g°C) is heated from 25.0°C to 245°C?

    a) 1.45×10^3 J   b) 134 J   c) 4.75×10^4 J   d) No correct answer given.

19. Homogenized whole milk contains 4% butterfat by mass. If 1g of butterfat supplies 9 calories, how many calories are contained in a 250 ml glass of milk (d=0.8 g/ml)?

    a) 10 cal   b) 72 cal   c) 250 cal   d) 40 cal

20. A 100.0-g sample of iron at 75.0°C is added to 200.0 g of H_2O at 20.0°C. The temperature of water rises to 22.9°C. Assuming that the calorimeter is a perfect insulator, what is the specific heat of the iron? Specific heat of H_2O is 4.184 J/g°C

    a) 0.47 J/g°C   b) 0.24 J/g°C   c) 4.7 J/g°C   d) 0.74 J/g°C

21. Consider the specific heats of the following substances.

Substance	Specific Heat
water	4.184 J/g °C
aluminum	0.900 J/g °C
copper	0.385 J/g °C
gold	0.131 J/g °C

If 40.0 kJ of energy is absorbed by 100.0 g of each substance at 25°C, which substance will have the lowest final temperature?

    a) copper   b) aluminum   c) gold   d) water

22. A 12.5 ml NaOH solution (d = 1.10 g/ml, specific heat = 4.10 J/g°C) absorbs 1.674 kJ. The temperature of the NaOH solution will increase by

    a) 14.8°C   b) 18.0°C   c) 9.2°C   d) 14.2°C

23. If 44 g of anthracite coal raise the temperature of 4.0 L of water from 20.0°C to 100.0°C, how much heat is given by 1 g of coal?

    a) 7000. cal   b) 2450. cal   c) 2700. cal   d) 3500. cal

24. Which of the following illustrates a situation in which essentially all the energy is potential, and none of it is kinetic?

    a) a stretched rubber band   b) a book sitting on a shelf
    c) a gallon of gasoline   d) All of the above.

25. The kinetic energy of an object depends on

    a) mass   b) density   c) volume   d) position

26. Which of these illustrates an increase in potential energy?

    a) a firecracker exploding   b) a person climbing a set of stairs
    c) a wind-up toy winding down   d) an apple dropping off a tree

27. Which of these materials will require the greatest amount of energy to warm from 30°C to 40°C?

    a) water   b) copper   c) gold   d) lead

28. When 9.44 g of calcium are heated in air, 13.22 g of calcium oxide are formed. The percent by mass of oxygen in the compound is

    a) 28.6%   b) 40.0%   c) 71.4%   d) 13.2%

29. Mercury (II) sulfide, HgS, contains 86.2% mercury by mass. The grams of HgS that can be made from 30.0 of mercury are

    a) 2.59×10^3 g   b) 2.87 g   c) 25.9 g   d) 34.8 g

30. The changing of liquid water to ice is known as a

    a) chemical change
    b) heterogeneous change
    c) homogeneous change
    d) physical change

31. Which has the highest specific heat?

    a) Ice   b) Lead   c) Water   d) Aluminum

32. When 20.0 g of mercury is heated from 10.0°C to 20.0°C, 27.6 J of energy are absorbed. What is the specific heat of mercury?

    a) 0.725 J/g°C   b) 0.138 J/g°C   c) 2.76 J/g°C   d) No correct answer given.

33. The diagram shows a pendulum bob as it swings. At which point does it have maximum potential energy?

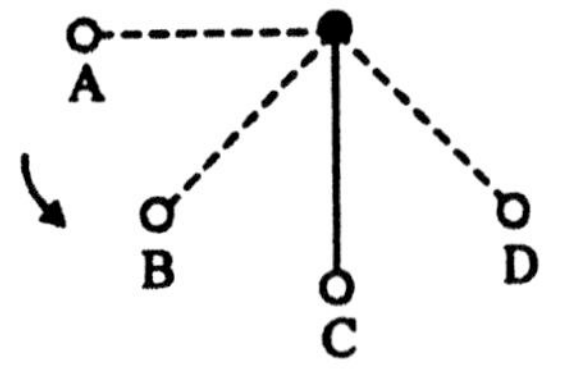

    a) A   b) B   c) C   d) D

34. In the same diagram, at which point does the bob have maximum kinetic energy?

    a) A   b) B   c) C   d) D

35. A shiny silver strip of metal is heated in a hot flame. It turns powdery and white. When the flame is removed, it remains powdery white. What can you conclude from these observations?

      a) All metals oxidize with heat.
      b) The strip has undergone a physical change.
      c) All physical changes require heat.
      d) The strip has undergone a chemical change.

CHAPTER 5.   Early Atomic Theory and Structure

True — False (choose one)

1.   Dalton's atomic theory states that electrons exist in orbitals which have a specific energy and shape.

2.   Dalton's atomic theory states that: (1) atoms of the same element are alike in mass and size, and (2) compounds are formed by the union of two or more atoms of different elements.

3.   The mass of an atom increases or decreases during a chemical reaction.

4.   The emissions generated in a Crookes tube are produced by electrons.

5.   The space outside the nucleus is mostly empty.

6.   The subatomic particles, protons, neutrons, and electrons, are located in the nucleus of the atom.

7.   An electron has 1/1837 the mass of a proton.

8.   The work that led to the proposal that the atom has a nucleus was Rutherford's experiment in which he bombarded gold-foil with alpha particles.

9.   A negative ion is considered to be a neutral atom which has lost one or more electrons.

10.   A positive ion is considered to be a neutral atom which has lost one or more electrons.

11.   An atom of sulfur, $_{16}^{32}S$, contains 16 protons in its nucleus.

12.   The atomic number of an element is the total number of protons and neutrons in the nucleus of an atom of that element.

13.   Most atoms have the same number of protons and neutrons in their nuclei.

14.   The international standard for atomic mass is based on the mass of $_{8}^{16}O$.

15.   A neutral atom contains equal numbers of protons and electrons.

16.   A hydrogen atom contains one proton and one neutron.

17.   The mass of a hydrogen atom is twice the mass of a proton.

18.   The three isotopes of hydrogen are called protium, deuterium, and tritium.

19.   A deuterium nucleus contains one proton and two neutrons.

20.   An atom of $_{15}^{31}P$ contains 15 protons, 16 neutrons and 31 electrons.

21.   One atomic mass unit is exactly equal to 1/12 the mass of a carbon-12 atom.

22.   The mass number is the total number of protons and neutrons in the nucleus of an atom.

23. An atom of $^{35}_{17}\text{Cl}$ contains 17 protons, 17 electrons and 35 neutrons.

24. In the isotope $^{75}_{33}\text{As}$, A = 33 and Z = 75.

25. The atomic mass of an element is the relative average mass of the isotopes of that element compared to the mass of carbon-12.

26. $^{235}_{92}\text{U}$ and $^{238}_{92}\text{U}$ are isotopes.

27. An atom of $^{18}_{8}\text{O}$ contains two more neutrons than an atom of $^{16}_{8}\text{O}$.

28. An electrically charged atom is called an isotope.

29. Isotopes of hydrogen differ in the number of neutrons in the nucleus.

30. A compound contains two or more elements in a definite composition by mass.

31. John Dalton did *not* propose the existence of positive and negative charges within the atom.

32. J.J. Thomson did *not* propose an organization to the location of electrons within the atom.

33. Ernest Rutherford did *not* propose a location for the positive charge in an atom.

34. The atomic number of an element is always greater than its mass number.

35. The ratio of protons to neutrons within an atom is always 1:1.

36. The proton was discovered by James Chadwick in 1932.

37. The Law of Definite Composition states that a compound contains two or more elements combined in a definite proportion by mass.

38. One atomic mass unit is defined as one-twelfth the mass of a carbon-12 atom.

39. The proton and neutron have approximately equal charge.

40. $^{35}_{17}\text{Cl}$ and $^{37}_{17}\text{Cl}$ are isotopes of chlorine.

41. In the isotope $^{112}_{47}\text{Ag}$, $Z = 112$ and $A = 47$.

42. All the isotopes of an element have the same number of electrons.

43. The compounds HgO and Hg_2O illustrate the Law of Multiple Proportions.

44. Electric charge can be either positive or negative.

45. Positive ions are called *cations*; negative ions are called *anions*.

46. All isotopes of the same element have the same number of neutrons.

47. The element represented by $^{51}_{21}\text{X}$ is antimony, Sb.

48. If a neutral atom has an atomic number of 29 and a mass number of 61, then the atom must contain 90 neutrons.

49. Atom X has 9 protons, 9 electrons, and 10 neutrons. Atom Y has 10 protons, 10 electrons, and 9 neutrons. From this we can conclude that atoms X and Y have the same mass number.

50. The nuclear charge in a Ca^{2+} ion is +18.

Multiple Choice (**choose the best answer**)

1. Which of the following is *not* part of Dalton's Atomic Theory?

    a) All elements are composed of atoms.
    b) The atoms are always moving.
    c) Atoms of the same element are alike in mass and size.
    d) Atoms of two elements can combine in different ratios to form more than one compound.

2. Which of the following is false?

    a) The proton has a charge of +1.
    b) The electron has a relative mass of 1/1837 amu.
    c) The proton has a relative mass of 1 amu.
    d) The neutron has a charge of +1.

3. Which subatomic particle has a relative mass of approximately one mass unit and a positive charge of +1?

    a) neutron   b) proton   c) electron   d) alpha particle

4. Atoms of the same atomic number but different atomic masses are called

    a) isomers   b) orbitals   c) neutrons   d) isotopes

5. An atom containing 19 protons, 20 neutrons, and 19 electrons has a mass number of

    a) 19   b) 20   c) 39   d) 58

6. Atoms of $^{38}_{20}Ca$ and $^{40}_{20}Ca$ differ with respect to number of

    a) protons   b) electrons   c) neutrons   d) orbitals

7. Two atoms with the same number of neutrons are called

    a) isotopes   b) deuterium   c) tritium   d) No correct answer given.

8. Isotopes always have the same

    a) mass number   b) number of electrons
    c) number of neutrons   d) atomic number

9. The atomic number of an element is related to its

   a) atomic mass   b) number of neutrons in the nucleus
   c) mass number   d) number of protons in the nucleus

10. The element with atomic number 53 contains

    a) 53 neutrons
    b) 53 protons
    c) 26 neutrons and 27 protons
    d) 26 protons and 27 neutrons

11. All atoms are electrically

    a) neutral   b) positive   c) negative   d) positive or negative depending on the atom

12. Which of the following statements is incorrect?

    a) ^{16}O and ^{18}O have the same number of protons.
    b) Ca^{2+} and Ar have the same number of electrons.
    c) ^{16}O and $^{16}O^{2-}$ have the same number of protons and neutrons.
    d) Isotopes of the same element have the same number of neutrons but different number of protons.

13. How many protons, electrons and neutrons are in $^{56}Fe^{3+}$?

	protons	electrons	neutrons
a)	26	26	30
b)	29	26	30
c)	26	23	30
d)	56	23	26

14. What is the mass of one oxygen atom?

    a) 2.66×10^{-23} g   b) 2.66×10^{23} c) 16.00 g   d) 1.66×10^{-24} g

15. Atom A has 5 protons and 6 neutrons; atom B has 6 protons and 5 neutrons. These atoms are

    a) isotopes
    b) isomers
    c) atoms of different elements
    d) identical in physical properties

16. The mass of one atom of an isotope is 9.746×10^{-23} amu. The atomic mass of this isotope is

    a) 5.870 amu   b) 16.18 amu   c) 58.69 amu   d) No correct answer given.

17. The two stable isotopes of boron have masses and abundance as follows: 10.0129 amu (19.91%) and 11.0129 amu (80.09%). What is the relative atomic mass of boron?

    a) 10.81 amu   b) 10.21 amu   c) 10.62 amu   d) 10.51 amu

18. Which of the following is an example of the Law of Multiple Proportions?

    a) a mixture of 30 g sulfur, 20 g iron, and 15 g of aluminum
    b) two different compounds of nitrogen and oxygen, with different amounts of each element in them
    c) a compound containing twice as much hydrogen as oxygen and twice as much oxygen as nitrogen
    d) two different physical forms of the same element

19. The Law of Definite Proportions pertains to

    a) compounds of elements   b) mixtures of compounds
    c) compounds of mixtures   d) mixtures of elements

20. The neutron was discovered in 1932 by

    a) Dalton   b) Rutherford   c) Thomson   d) Chadwick

21. The number of neutrons in an atom of $^{139}_{56}Ba$ is

    a) 56   b) 83   c) 139   d) No correct answer given.

22. The name of the isotope containing one proton and two neutrons is

    a) protium   b) tritium   c) deuterium   d) helium

23. Which pair of symbols represents isotopes?

    a) $^{23}_{11}Na$ and $^{23}_{12}Na$ b) $^{7}_{3}Li$ and $^{6}_{3}Li$ c) $^{63}_{29}Cu$ and $^{29}_{64}Cu$ d) $^{12}_{24}Mg$ and $^{12}_{26}Mg$

24. Two naturally occurring isotopes of an element have masses and abundance as follows: 54.00 amu (20.00%) and 56.00amu (80.00%). What is the relative atomic mass of the element?

    a) 54.20   b) 54.40   c) 54.80   d) 55.60

25. Naturally occurring copper consists of 2 isotopes ^{63}Cu (atomic mass 62.9 amu) and ^{65}Cu (atomic mass 64.9 amu). The average atomic mass of copper is 63.5 amu. What are the natural abundances of these 2 isotopes?

	^{63}Cu	^{65}Cu
a)	60%	40%
b)	30%	70%
c)	70%	30%
d)	50%	50%

26. The mass ratio of carbon to oxygen in carbon monoxide, CO, is 0.7506. What mass of carbon monoxide is formed when 3 g of carbon combine with oxygen?

    a) 7.00   b) 4.00 g   c) 7.50 g   d) 14.0 g

27. The number of neutrons in an atom of $^{108}_{47}Ag$ is

   a) 47   b) 108   c) 155   d) No correct answer given.

28. The number of electrons in an atom of $^{27}_{13}Al$ is

   a) 13   b) 14   c) 27   d) 40

29. The number of electrons in the nucleus of an atom of $^{24}_{12}Mg$ is

   a) 12   b) 24   c) 36   d) No correct answer given.

30. Which experiment led to the notion that the atom contains an extremely small, positively charged nucleus?

   a) Millikan's oil drop experiment
   b) Rutherford's scattering experiment
   c) Thomson's cathode ray tube experiment
   d) Dalton's atomic theory

31. Which of the following statements is false?

   a) Every atom contains electrons, protons, and neutrons.
   b) Electrons and protons are electrically charged.
   c) The mass of a proton is about the same as the mass of a neutron.
   d) All are correct.

32. If N is the number of neutrons in a given nucleus, which one of the following relations is correct?

   a) $N = A + Z$ b) $Z = A + N$ c) $N = A - Z$ d) $A = N - Z$

33. The atomic mass listed on the Periodic Table for each element is

   a) the mass of the most abundant isotopes of the element
   b) the weighted average of the masses of the naturally-occurring isotopes of the element
   c) the arithmetic average of the mass of the naturally-occurring isotopes of the element
   d) the ratio of the mass of atom of the element to the mass of one hydrogen atom

34. A particular ion has as its formula $^{90}_{38}Sr^{2+}$. The number of electrons found in one of these ions is

   a) 90   b) 88   c) 36   d) 2

35. Ca-40, K-39, and Sc-41 all have the same

   a) atomic mass   b) atomic number   c) number of electrons   d) number of neutrons

36. The mass of one atom of iron is 9.274×10^{-23} g. What percent of this mass comes from electrons?

   a) 5%   b) 0.2765%   c) 0.02554%   d) 0.00225%

Matching

From the list of terms given, choose the one that correctly identifies each phrase.

List of terms: (a) Law of Multiple Proportions; (b) James Chadwick; (c) Law of Definite Composition; (d) J. J. Thomson; (e) atom; (f) atomic number; (g) atomic mass unit; (h) prime numbers; (i) anion; (j) atomic mass; (k) number of neutrons; (1) Rutherford; (m) nucleus; (n) isotopes; (o) electron; (p) cation.

_____ **1.** He discovered the neutron.

_____ **2.** 1/12 the mass of a carbon-12 atom.

_____ **3.** He showed that the proton is a particle and calculated its mass to be about 1837 times that of an electron.

_____ **4.** The average relative mass of the isotopes of an element referred to the atomic mass of carbon-12.

_____ **5.** The compounds FeO, Fe_2O_3, and Fe_3O_4 are examples of what law?

_____ **6.** A negatively charged ion.

_____ **7.** The same as the number of protons in the nucleus.

_____ **8.** Atoms that have the same atomic number but different mass numbers.

_____ **9.** A particle with a relative mass of $\dfrac{1}{1837}$ amu.

_____ **10.** The part of the atom containing the protons.

CHAPTER 6.   Nomenclature of Inorganic Compounds

True — False (choose one)

1. The compound formed between Sn^{4+} and SO_4^{2-} is Sn_2SO_4.

2. The compound formed between NH_4^+ and PO_4^{3-} is $(NH_4)_3PO_4$.

3. The common name for HNO_3 is muriatic acid.

4. The formula for mercury (I) oxide is Hg_2O.

5. The prefixes tetra and hexa mean four and six, respectively.

6. When the name of an oxy-acid ends in *ic*, the corresponding salt name will end in *ite*.

7. The name for Cl_2O_7 is dichlorine octoxide.

8. The name for $SnCr_2O_7$ is tin(II) dichromate.

9. The compound formed between Fe^{3+} and CO_3^{2-} is $Fe2(CO_3)_3$.

10. The formula for ammonium chloride is NH_4Cl, and that for ammonium sulfide is NH_4S.

11. The name for H_2S is hydrosulfuric acid, and that for HBr is hydrobromic acid.

12. The formulas for sodium hydroxide, potassium hydroxide, and calcium hydroxide are NaOH, KOH, and $Ca(OH)_2$, respectively.

13. Six molecules of Al_2O_3 contain more atoms of oxygen than 12 molecules of water.

14. When a chemical formula is written, the more metallic element is written first.

15. Sulfur hexafluoride would be symbolized as SF_5.

16. Atoms form ions on their own when forming compounds.

17. The charge of an ion can often be predicted from the position of the element on the periodic table.

18. Transition metals can form multiple charges.

19. The oxidation number of oxygen in peroxides is -2.

20. The compound formed from NH_4^+ and SO_4^{2-} is $(NH_4)_2SO_4$.

21. The nitrite ion has three oxygen atoms and the nitrate ion has four oxygen atoms.

22. If the name of an acid ends in *ous*, the corresponding salt name will end in *ate*.

23. The formula for cane or beet sugar is $C_6H_{12}O_6$.

24. The name for Cl_2O_3 is dichloroheptoxide.

25. The name for $NaHCO_3$ is sodium hydrogen carbon trioxide.

26. The name for KHC_2O_4 is potassium hydrogendicarbontetraoxide.

27. A chemical formula is a shorthand expression for a chemical reaction.

28. The formula for copper (II) sulfate is $CuSO_4$.

29. The formula for mercury (I) carbonate is $HgCO_3$.

30. The formula for phosphorus triiodide is PI_3.

31. The formula for calcium acetate is $Ca(C_2H_3O_2)_2$.

32. The formula for hypochlorous acid is $HClO$.

33. The formula for dichlorine heptoxide is Cl_2O_7.

34. The formula for magnesium iodide is MgI.

35. The formula for sulfurous acid is H_2SO_3.

36. The formula for potassium manganate is $KMnO_4$.

37. The formula for lead (II) chromate is $PbCrO_4$.

38. The formula for ammonium bicarbonate is NH_4HCO_3.

39. The formula for iron (II) phosphate is $FePO_4$.

40. The formula for calcium hydrogen sulfate is $CaHSO_4$.

41. The formula for mercury (II) sulfate is $HgSO_4$.

42. The formula for sodium hypochlorite is $NaClO$.

43. The formula for sodium dichromate is $Na_2Cr_2O_7$.

44. The formula for cadmium cyanide is $Cd(CN)_2$.

45. The formula for bismuth (III) oxide is Bi_3O_2.

46. The formula for carbonic acid is H_2CO_3.

47. The formula for silver oxide is Ag_2O.

48. The formula for tin (II) fluoride is TiF_2.

49. The formula for carbon monoxide is CO.

50. The formula for phosphoric acid is H_3PO_3.

51. The formula for sodium bromate is Na_2BrO_3.

52. The formula for hydrosulfuric acid is H_2S.

53. The formula for potassium hydroxide is POH.

54. The formula for zinc sulfate is $ZnSO_3$.

55. The formula for sulfur trioxide is SO_3.

56. The formula for tin (IV) nitrate is $Sn(NO_3)_4$.

57. The formula for ferrous sulfate is $FeSO_4$.

58. The formula for chloric acid is HCl.

59. The formula for aluminum sulfide is Al_2S_3.

60. The formula for cobalt (II) chloride is $CoCl_2$.

61. The formula for acetic acid is $HC_2H_3O_2$.

62. The formula for zinc oxide is ZnO_2.

63. The formula for stannous fluoride is SnF_2.

64. In the ionic compound $K_2Cr_2O_7$, the charge on each chromium atom is +12.

65. Common ions which carry a -3 charge include PO_4^{3-}, PO_3^{3-}, and CO_3^{3-}.

66. Sulfuric acid can also be called hydrogen sulfate.

67. Nitric acid can also be called hydrogen nitrate.

68. In the compound CaO, the ionic charge on calcium is +1.

69. Sodium sulfide, sodium sulfite, and sodium sulfate each contain the same three elements.

70. The sulfurous acid has less oxygen atoms than the sulfuric acid.

Multiple Choice (choose the best answer)

1. Which compound's name ends in *-ate*?

    a) $HClO_2$ b) $NaNO_2$ c) $MgSO_4$ d) $Al_2(SO_3)_3$

2. The correct formula for iron(III) sulfide is

    a) FeS b) Fe_2S_3 c) FeS_3 d) Fe_3S

3. The formula for a compound between Ba and O is most likely to be

    a) Ba_2O b) BaO c) BaO_2 d) Ba_2O_3

4. The formula for carbon tetrachloride is

    a) CCl_4 b) C_2Cl_4 c) C_2Cl_6 d) No correct formula given.

5. Which formula represents the substance commonly known as lime?

    a) CaO b) $NaOH$ c) $CaCO_3$ d) Al_2O_3

6. Which formula represents the substance commonly known as pyrite or fool's gold?

    a) PbO b) FeS_2 c) PbS d) Hg

7. The name for $Pb(NO_2)_2$ is

    a) lead (II) nitrate   b) lead(IV) nitrate   c) lead(II) nitrite   d) lead(III) nitrate

8. The name for $HClO_3$ is

    a) hydrochloric acid   b) hydrogen chloride oxide   c) chlorous acid   d) chloric acid

9. Which formula is hypochlorous acid?

    a) $HClO_4$ b) H_2ClO_2 c) $HClO$ d) $HClO_3$

10. All of the following species acquire a -2 in ionized form except

    a) chloride   b) sulfide   c) carbonate   d) sulfate

11. Which of these statements regarding a neutral ionic compound is not true?

    a) All charges in the compounds must add to 0.
    b) The negative ion is always written before the positive ion.
    c) Ionic compounds must always contain metals and nonmetals.
    d) All ionic compounds include at least two elements.

12. Which one of the following combinations is incorrect?

    a) FeO, iron (II) oxide
    b) CCl_4, carbon (IV) chloride
    c) CuO, copper (II) oxide
    d) $CrCl_3$, chromium (III) chloride

13.   Which one of the following combinations is incorrect?

    a) H_2SO_4, sulfuric acid   b) $KMnO_4$, potassium permanganate
    c) $CuSO_4$, copper sulfite   d) KIO_3, potassium iodate

14.   The correct formula for iodic acid is

    a) HI   b) H_2IO_4 c) HIO_3 d) HIO_4

15.   The name for $Na_2C_2O_4$ is

    a) sodium oxalate   b) sodium carbonate
    c) sodium dicarbon tetraoxide   d) sodium (I) carbonate

16.   What is the charge on the manganese in $KMnO_4$?

    a) +1   b) +7   c) +6   c) −7

17.   Which one of the following is the sulfate ion?

    a) S^{2-} b) HSO_4^- c) SO_4^{2-} d) SO_3^{3-}

18.   The parent acid of the nitrate ion is

    a) nitric acid   b) nitrous acid   c) acid nitrate   d) hydronitric acid

19.   The parent acid of the chlorate ion is

    a) perchloric acid   b) chloric acid   c) hypochlorous acid   d) chlorous acid

20.   Element X combines with phosphate ion to form $X_3(PO_4)_2$. Element X is

    a) NH_4^+ b) Al^{3+} c) Ca^{2+} d) N^{3-}

Matching I

Match each anion with its formula.

List of terms: (a) CO_3^{2-}; (b) HCO_3^-; (c) S^{2-}; (d) O^{2-}; (e) SO_3^{2-}; (f) PO_4^{3-}; (g) NO_2^-; (h) BrO_2^-; (i) BrO_3^-; (j) SO_4^{2-}; (k) $C_2H_3O_2^-$; (l) Cl^-; (m) ClO_4^-; (n) ClO_3^-; (o) NO_3^-; (p) CrO_4^{2-}; (q) $Cr_2O_7^{2-}$; (r) OH^-; (s) AsO_4^{3-}; (t) H^-

_____ 1.   Chloride

_____ 2.   Bromite

_____ 3.   Nitrate

_____ 4.   Sulfite

_____ 5.   Hydrogen carbonate

_____ 6.   Chlorate

_____ 7.   Chromate

_____ 8.   Phosphate

_____ 9.   Oxide

_____ 10.   Hydroxide

Matching II

Match each cation with its formula.

List of terms: (a) AS^{3+}; (b) Sn^{2+}; (c) I^{3+}; (d) La^{3+}; (e) Mg^{2+}; (f) Ca^{2+}; (g) So^+; (h) Pb^{2+}; (i) Zn^{2+}; (j) K^+; (k) CU^+; (l) CU^{2+}; (m) Ag^+; (n) Ar^{3+}; (o) Na^+; (p) Ma^{2+}; (q) Fe^{3+}; (r) Si^{4+}; (s) Zr^{4+}; (t) P^{3+}

_____ 1.   Arsenic(III)

_____ 2.   Lead(II)

_____ 3.   Calcium

_____ 4.   Iron(III)

_____ 5.   Silver

_____ 6.   Sodium

_____ 7.   Magnesium

_____ 8.   Zinc

_____ 9.   Copper(I)

_____ 10.   Potassium

Matching III

Match each acid with its correct formula.

List of terms: (a) $H_2C_2O_4$; (b) HCl; (c) HNO_3; (d) H_3PO_3; (e) $HC_2H_3O_2$; (f) H_2SO_3; (g) H_3PO_4; (h) HNO_2; (i) $HClO$; (j) $HClO_3$; (k) H_3N; (l) $HClO_4$; (m) H_2S; (n) H_2SO_4; (o) H_2CO_3

_____ 1.   Hydrochloric acid

_____ 2.   Hydrosulfuric acid

_____ 3.   Nitric acid

_____ 4.   Acetic acid

_____ 5.   Sulfuric acid

_____ 6.   Phosphoric acid

_____ 7.   Nitrous acid

_____ 8.   Chloric acid

_____ 9.   Perchloric acid

_____ 10.   Phosphorous acid

Completion

Write the formulas and names of the compounds which can be formed from the ions given.

_____ 1. H^+ CO_3^{2-}

_____ 2. Fe^{2+} NO_3^-

_____ 3. Al^{3+} Cl^-

_____ 4. Ca^{2+} PO_4^{3-}

_____ 5. NH_4^+ SO_4^{2-}

_____ 6. Hg^+ SO_3^{2-}

_____ 7. Mg^{2+} ClO^-

_____ 8. K^+ ClO_3^-

_____ 9. Na^+ ClO_4^-

_____ 10. Cu^+ O^{2-}

_____ **11.** Ni^{2+} $C_2H_3O_2^-$ _____ **16.** Li^+ I^-

_____ **12.** Sn^{4+} CrO_4^{2-} _____ **17.** Fe^{3+} HCO_3^-

_____ **13.** Ba^{2+} $Cr_2O_7^{2-}$ _____ **18.** Cd^{2+} CN^-

_____ **14.** K^+ MnO_4^- _____ **19.** Zn^{2+} OH^-

_____ **15.** Na^+ S^{2-} _____ **20.** Ag^+ Br^-

CHAPTER 7.   Quantitative Composition of Compounds

True — False (choose one)

1.   A mole is an amount of substance that contains the same number of particles as there are atoms in exactly 24 g of carbon-12.

2.   The atomic mass and the molar mass of ^{12}C are 12 a.m.u. and 12 g/mol respectively.

3.   Carbon-12 and carbon-14 are the two references for atomic mass.

4.   A mole contains Avogadro's number of formula units.

5.   Avogadro's number is 6.022.

6.   A mole of sulfur atoms weighs 6.022×10^{23} grams.

7.   A mole of oxygen molecules (O_2) weighs the same as a mole of oxygen atoms.

8.   3.0 atomic masses of aluminum are equal to 9.0 moles of aluminum.

9.   One mole of ammonia, NH_3, contains 2.409×10^{23} atoms.

10.   Two moles of O_3 contain the same number of atoms as one mole of C_2H_4.

11. 6.022×10^{23} silver atoms is one mole of silver.

12.   0.5 mol of H_2 contains 6.022×10^{23} hydrogen atoms.

13.   One mole of $Ca(NO_3)_2$ contains one mole of Ca atoms, one mole of N atoms and 3 moles of O atoms.

14.   A sample of Ca weighing 40 grams contains the same number of moles as a sample of Co weighing 40 grams.

15.   One mole of KNO_2 contains more oxygen atoms than one mole of $NaNO_2$.

16.   10.0 g of KNO_2 contains more grams of oxygen than 10.0 g of $NaNO_2$.

17.   10.0 g of KNO_2 contains more oxygen atoms than 10.0 g of $NaNO_2$.

18.   A mole of sodium sulfite, Na_2SO_3, and a mole of sodium sulfate, Na_2SO_4, contain the same number of sodium atoms.

19.   The number of atoms in 9.0 g of H_2O is 9.0×10^{23} atoms.

20.   5.00 moles of H_2O contain twice as many hydrogen atoms as oxygen atoms.

21.   5.00 grams of H_2O contain twice as many hydrogen atoms as oxygen atoms.

22.   The molar mass of a substance is the total mass of all the atoms in the chemical formula of that substance.

23. A compound with an empirical formula of CH_2 can have a molar mass of 112.

24. The molar mass of $CaCl_2$ is 111.1g/mol.

25. One-fourth of a mole of oxygen gas weighs 4.0 g.

26. The sum of the mass percents of each element in a compound equals 100%.

27. The smallest ratio of the atoms that are present in a formula of a compound represents the simplest formula of that compound.

28. The empirical formula and molecular formula of a compound always have the same percent composition.

29. Sometimes the empirical and molecular formulas of a compound are identical.

30. The molecular formula always represents the total number of atoms of each element present in one molecule of a compound.

31. The only information needed to calculate the empirical formula of a compound is which elements are in the compound and their atomic masses.

32. The subscripts in a molecular formula represent the number of atoms of each element in one molecule of the compound and also represent the number of moles of each element in one mole of the compound.

33. If two separate substances are found to have identical percent compositions then the two substances must be identical.

34. It is possible for two different compounds to have identical empirical formulas.

35. The mass of 1 atom of hydrogen, H_2, is 2.018 g/6.022 x 10^{23} atoms.

36. The mass of 1 molecule of aspirin, $C_9H_8O_4$, can be obtained by dividing the molar mass of aspirin by Avogadro's number.

37. The empirical formula for aspirin, $C_9H_8O_4$, is $C_4H_4O_2$.

38. The compound glucose, $C_6H_{12}O_6$, has the same empirical formula as the compound sucrose $C_{12}H_{22}O_{11}$.

39. The symbol (or formula) $Fe[Fe(CN)_6]_4$ shows 10 atoms of C.

40. A mole contains Avogadro's number of atoms, molecules, or formula units.

41. One mole of chlorine molecules contains 2 mol of chlorine atoms.

42. The mass of 2 mol of hydrogen molecules is 4.032 g.

43. $CaCl_2$ has a higher percentage of chlorine than $MgCl_2$.

44. A compound has an empirical formula of C_2H_2O and a molar mass of 168.0. The molecular formula is $C_6H_6O_3$.

45. One mole of $HC_2H_3O_2$ has a mass of 60.05 g.

46. If the molecular formula and empirical formula of a compound are not the same, the empirical formula will be an integral multiple of the molecular formula.

47. The number of sulfur atoms is the same in 1 g of Na_2SO_4 as in 1 g of K_2SO_4.

48. The molar mass of a compound is the sum of the molar masses of all the atoms in the formula of the compound.

49. The formula for sucrose, $C_{12}H_{22}O_{11}$, is empirical.

50. Nitrogen dioxide and dinitrogen tetroxide have the same empirical formula.

51. Grams measure the mass of particles; moles measure the number of particles.

52. A compound contains 4×10^{23} atoms of calcium, 21.3 g of sulfur, and 1. 33 moles of oxygen. Its empirical formula is $CaSO_4$.

Multiple Choice (choose the best answer)

1. Which statement is incorrect; one mole equals:

a) 6.022×10^{23} molecules or formula units   b) 1 molar mass of a compound
c) 1 atomic mass of a monatomic element   d) 6.022×10^{23} grams of a compound

2. Which pair of compounds would have the same percent composition?

a) Hg_2Cl_2 and $HgCl$ b) $NaHCO_3$ and Na_2CO_3
c) K_2CrO_4 and $K_2Cr_2O_7$ d) H_2O and H_2O_2

3. One mole of hydrogen atoms contams

a) 2.016 g   b) 6.022×10^{23} atoms   c) 1 atom   d) 12 g of carbon-12

4. One atom of magnesium weighs

a) 24.31 g   b) 6.022×10^{23} amu   c) 54.94 g   d) 4.04×10^{-23} g

5. One mole of oxygen atoms contains

a) 32.00 g   b) 1.00 g   c) 6.022×10^{23} atoms   d) 16 atoms

6. 7.0 g of nitrogen (N_2) contains

a) 7.0 atoms of N_2 b) 3.0×10^{23} atoms of N_2
c) 0.25 mole of N_2 d) 7.0 atomic masses of N_2

7. Which statement is false regarding a mole?

   a) One mole of H_2O contains the same number of molecules as one mole of O_2
   b) One mole of H_2O contains 2 atoms of H
   c) 12 g of carbon-12 contains one mole of ^{12}C
   d) A mole is a SI unit for counting atoms and molecules

8. All of the following contain the same amount of particles except

   a) 1 mole of H_2
   b) 32 g of O_2
   c) a container with 16 g of O_2 and 0.5 mole of H_2
   d) 1 mole of NaCl dissolved in water

9. You spend 1 mole of $ at a rate of 1 million $ every second during your lifetime (90 years). What percent of the original money did you spend?

   a) all of it
   b) 10%
   c) less than 0.00001%
   d) 1%

10. One atom of chlorine weighs

    a) 35.45 g   b) 17.73 g   c) 6.022×10^{23} g   d) 5.90×10^{-23} g

11. 2.7×10^{24} molecules of CH_4

    a) weigh 72 g   b) is 0. 17 mole   c) is 0.45 mole   d) weigh 23 g

12. The molar mass of $MgCl_2$ is

    a) 95.21 g   b) 59.76 g   c) 119.5 g   d) 125.8 g

13. The formula mass of C_3H_8O is

    a) 29.02   b) 53.03   c) 66.08   d) No correct answer given.

14. How many moles are contained in 2.54 g Cu?

    a) 0.0400 mol   b) 25.0 mol   c) 161 mol   d) No correct answer given.

15. How many moles are contained in 14.2 g $BaSO_4$?

    a) 14.2 mol   b) 0.706 mol   c) 0.608 mol   d) 2.36×10^{-23} mol

16. How many moles of atoms are contained in 12.4 g CH_4?

    a) 0.773   b) 7.47×10^{24} C) 198   d) 3.87

17. Which of the following contains the largest number of moles?

    a) $2.0\,g\,CO_2$ b) $2.0\,g\,O_2$ c) $2.0\,g\,C_2H_6$ d) $2.0\,g\,HCl$

18. How many grams of H_2O will contain a total number of atoms equal to Avogadro's number?

    a) $18.02\,g$ b) $9.010\,g$ c) $6.000\,g$ d) $3.000\,g$

19. What is the atomic mass of an element if one atom has a mass of 2.107×10^{-22} g?

    a) 253.8 b) 126.9 c) 63.45 d) 3.042×10^{23}

20. Which of the following statements is incorrect?

    a) The number of hydrogen atoms in 1 g of H_2O is equal to the number of hydrogen atoms in 1 g of H_2O_2.

    b) The number of hydrogen atoms in 1 mol of H_2O is equal to the number of hydrogen atoms in 1 mol of H_2O_2.

    c) The number of oxygen atoms in 1 mole of H_2O_2 is twice the number of oxygen atoms in 1 mole of H_2O.

    d) The total number of atoms in 1 mole of H_2O_2 is larger than the total number of atoms in 1 mole of H_2O.

21. The number of oxygen atoms in 7.8 g of O_2 is

    a) 2.93×10^{23} b) 1.47×10^{23} c) 5.86×10^{23} d) 0.488

22. The number of nitrate ions in 56.2 g of magnesium nitrate is

    a) 2.28×10^{23} b) 4.56×10^{23} c) 3.92×10^{23} d) 7.84×10^{23}

23. How many grams of calcium nitrate are contained in 2.52 mol of $Ca(NO_3)_2$?

    a) $69.1\,g$ b) $414\,g$ c) $74.1\,g$ d) $257\,g$

24. How many grams of sodium are contained in 0.255 mol of Na?

    a) $23.0\,g$ b) $90.2\,g$ c) $5.86\,g$ d) $0.0111\,g$

25. How many grams of silver are contained in 50.0 g of $AgNO_3$?

    a) $31.8\,g\,Ag$ b) $0.284\,g\,Ag$ c) $107.9\,g\,Ag$ d) $42.5\,g\,Ag$

26. How many grams of silver are contained in 4.52 mol of $AgNO_3$?

    a) $23.9\,g\,Ag$ b) $37.6\,g\,Ag$ c) $488\,g\,Ag$ d) $768\,g\,Ag$

27. What is the percent composition of Zinc chloride, $ZnCl_2$?

a) 52.02% Zn, 47.97% Cl   b) 64.86% Zn, 35.17% Cl
c) 47.97% Zn, 52.02% Cl   d) No correct answer given.

28. The percent composition of aluminum hydroxide, $Al(OH)_3$ is

a) 34.60% Al, 61.54% O, 3.877% H   b) 61.35% Al, 36.38% O, 2.27% H
c) 24.60% Al, 61.54% O, 13.86% H   d) No correct answer given.

29. Which compound has the highest percentage of calcium?

a) CaF_2 b) $CaCl_2$ c) $CaBr_2$ d) CaI_2

30. Which compound has the lowest percentage of chlorine?

a) $HClO$ b) $LiClO_2$ c) $NaClO_3$ d) $KClO_4$

31. A 7.33 g sample of lanthanum, La, combined with oxygen to give 10.33 g of an oxide. The percent composition of this oxide is

a) 59.1% La, 40.9% O   b) 24.4% La, 75.6% O
c) 71.0% La, 29.0% O   d) No correct answer given.

32. The empirical formula of the compound whose composition is 39.7% K, 27.8% Mn, and 32.5% O is

a) $KMnO_3$ b) $K_4Mn_3O_3$ c) K_2MnO_4 d) $KmnO_4$

33. An oxide of manganese has the following composition: 53.4% Mn and 46.6% O. The empirical formula of the oxide is

a) MnO_2 b) MnO_3 c) Mn_2O_3 d) MnO

34. Which one of the following could be an empirical formula?

a) C_3H_8 b) $C_6H_6O_6$ c) $C_2H_2Cl_2$ d) C_2H_6

35. Which one of the following could not be an empirical formula?

a) CH_2 b) CH_4 c) C_2H_5 d) C_2H_6

36. The composition of red lead oxide is 90.7% Pb and 9.33% O. The empirical formula of this oxide is

a) PbO b) PbO_2 c) Pb_2O_3 d) Pb_3O_4

37. What is the emprical formula of a compound that has 29.1% Na, 40.6% S, and 30.3% O?

a) Na_2SO_3 b) Na_2SO_4 c) $Na_2S_2O_3$ d) $Na_2S_4O_6$

38. What is the molecular formula of a compound which has an empirical formula of CH_2 amd a molar mass of 126.2?

      a) C_9H_{18} b) $C_{10}H_{20}$ c) C_8H_{30} d) $C_{12}H_{24}$

39. A compound contains 29.9% C, 57.8% Cl, and 13.0% O. The molar mass is 246.0. The molecular formula is

      a) $C_6Cl_4O_2$ b) $C_3Cl_5O_2$ c) $C_2Cl_4O_5$ d) $C_9Cl_3O_2$

40. A compound contains 54.5% C, 9.09% H, and 36.4% O. The molar mass is 88.10. The molecular formula is

      a) $C_2H_4O_2$ b) $C_3H_8O_2$ c) $C_4H_8O_2$ d) $C_5H_{12}O$

41. Which compound contains 35.0% N?

      a) NH_4NO_3 b) NH_4NO_2 c) NH_4Cl d) NH_2OH

42. The number of sulfur atoms in 96 g of sulfur is

      a) 1   b) 3   c) 6×10^{23} d) 1.8×10^{24}

43. The mass of one molecule of H_2O is

      a) 18 g   b) 3 g   c) 3×10^{-23} g   d) 1×10^{-23} g

44. 4.0 g of oxygen contain

      a) 1.5×10^{23} atoms of oxygen   b) 4.0 molar masses of oxygen
      c) 0.50 mol of oxygen   d) 6.022×10^{23} atoms of oxygen

45. One mole of hydrogen atoms contains

      a) 2.0 g of hydrogen   b) 6.022×10^{23} atoms of hydrogen
      c) 1 atom of hydrogen   d) 12 g of carbon-12

46. The mass of one atom of magnesium is

      a) 24.3 g   b) 54.9 g   c) 12.0 g   d) 4.035×10^{-23} g

47. Which of the following contains the largest number of moles?

      a) 1.0 g Li   b) 1.0 g Na   c) 1.0 g Al   d) 1.0 g Ag

48. How many moles of aluminum hydroxide are in one antacid tablet containing 400 mg of $Al(OH)_3$?

      a) 5.13×10^{-3} b) 0.400   c) 5.13   d) 9.09×10^{-3}

49. The molar mass of $Ba(NO_3)_2$ is

    a) 199.3 b) 261.3 c) 247.3 d) 167.3

50. What is the percent composition for a compound formed from 8.15 g of zinc and 2.00 g of oxygen?

    a) 80.3% Zn, 19.7% O   b) 80.3% O, 19.7% Zn
    c) 70.3% Zn, 29.7% O   d) 65.3% Zn, 34.7% O

51. A 3.056 g sample of vanadium combines with oxygen to form 5.456 g of vanadium oxide. The empirical formula of this compound is:

    a) VO_3 b) VO_5 c) V_2O_3 d) V_2O_5

52. Which of the following compounds contains the largest percentage of oxygen?

    a) SO_2 b) SO_3 c) N_2O_3 d) N_2O_5

53. The empirical formula of the compound containing 31.0% Ti and 69.0% Cl is

    a) $TiCl$ b) $TiCl_2$ c) $TiCl_3$ d) $TiCl_4$

54. What is the mass of 4.53 mol of Na_2SO_4?

    a) 142.1 g b) 644 g c) 31.4 g d) 3.19×10^{-3} g

55. The percent composition of Mg_3N_2 is

    a) 72.2% Mg, 27.8% N   b) 63.4% Mg, 36.6% N
    c) 83.9% Mg, 16.1% N   d) No correct answer given.

56. How many grams of oxygen are contained in 0.500 mol of Na_2SO_4?

    a) 16.0 g b) 32.0 g c) 64.0 g d) No correct answer given.

57. A sample of a compound contains 0.100 g of hydrogen and 4.20 g of nitrogen. The simplest formula for the compound is

    a) HN_2 b) HN_3 c) NH_3 d) NH_2

58. A compound is analyzed and found to contain 15.1% nitrogen, 34.5% oxygen, and the remainder cerium. What is the correct empirical formula for this compound?

    a) $Ce_2(NO_3)_2$ b) $Ce_2(NO_2)_3$ c) $Ce(NO_2)_3$ d) $Ce(NO_3)_2$

59. A particular organic compound is known to have the empirical formula C_2H_2O, and its molar mass is 504. Its molecular formula must be

    a) $C_{24}H_{36}O_{15}$ b) $C_{24}H_{24}O_{12}$ c) $C_{12}H_{72}O_{18}$ d) C_2H_2O

130

60. The element X forms a compound with bromine and its formula is XBr_2. When it is analyzed, the compound is found to have 26.9% X by mass. What is the atomic mass of X?

    a) 29.4 g/mol   b) 49.2 g/mol   c) 58.8 g/mol   d) 73.1 g/mol

61. Calcium sulfate is known to form a hydrated compound. When a 3.50 g sample of the hydrate is heated to dryness, the residue that remains has a mass of 2.77 grams. The formula of the hydrate is

    a) $CaSO_4 \cdot H_2O$ b) $CaSO_4 \cdot 2H_2O$ c) $CaSO_4 \cdot 5H_2O$ d) $CaSO_4 \cdot 8H_2O$

62. A 0.523 mole of a compound KXO_4 weighed 82.65 g. The element X is

    a) C   b) Mn   c) S   d) Cr

CHAPTER 8. Chemical Equations

True — False (choose one)

1. Subscripts following symbols in formulas are used to give the proper ratio of atoms in the formula.

2. In the expression 2 $AlCl_3$ the 2 is a coefficient and the 3 is a subscript.

3. The products are the substances produced by a chemical reaction.

4. A balanced equation has the same number of moles of reactants and products.

5. A balanced chemical equation represents a shorthand expression for a chemical and physical change.

6. In a balanced equation the same kinds of atoms are present on each side of the equation.

7. In a balanced equation the total number of atoms in the reactants equals the total number of atoms in the products.

8. In a balanced equation the total number of reactant atoms and molecules equals the total number of product atoms and molecules.

9. Balanced equations pertain only to those reactions which go to completion (100% conversion of reactants to products).

10. When the equation $Al(ClO_3)_3 \rightarrow AlCl_3 + O_2$ is balanced, the coefficient of O_2 will be 6.

11. In a balanced equation grams of reactants equal grams of products.

12. In a balanced equation the moles of reactants equal the moles of products.

13. If reactants are mixed in proportions other than those represented by the balanced equation then the products formed will not be in the proportions represented by the balanced equation.

14. One mole of ethyl alcohol, C_2H_6O, contains 2 moles of carbon atoms.

15. In the reaction 2 H_2 + O_2 → 2 H_2O, 100 molecules of H_2O are formed for every 50 molecules of O_2 that react.

16. In the reaction 2 Al + 6 HCl → 2 $AlCl_3$ + 3 H_2, two moles of H_2 will be formed when four moles of HCl are reacted.

17. When the equation Zn + HCl → $ZnCl_2$ + H_2 is balanced, the total number of moles of reactants and products will be 6.

18. In double displacement reactions, changes in the charges of some of the reactants occur.

19. In single displacement reactions, changes in the charges of the reactants occur.

20. All single displacement reactions involve reactions between metals and nonmetals.

21. The equation $NaCl + AgNO_3 \rightarrow AgCl(s) + NaNO_3$ represents a double displacement reaction.

22. The equation $2\,Na + Cl_2 \rightarrow 2\,NaCl$ represents a single displacement reaction.

23. The equation $BaCO_3 \xrightarrow{\Delta} BaO + CO_2$ represents a decomposition reaction.

24. The equation $Cl_2 + 2\,NaI \rightarrow 2\,NaCl + I_2$ represents a single displacement reaction.

25. The equation $Ca(OH)_2 + 2\,HCl \rightarrow CaCl_2 + 2\,H_2O$ represents a neutralization reaction.

26. Natural gas is primarily methane, CH_4.

27. The complete combustion of hydrocarbons produces CO_2 and H_2O as products.

28. A chemical change that liberates heat energy is said to be exothermic.

29. If the products of a reaction are at a lower energy level than the reactants were at the start of the reaction, then the reaction is endothermic.

30. A single displacement reaction results in just one product.

31. Chemical equations can be balanced by changing the subscripts of the formulas or by subtracting formulas of pure elements.

32. A decomposition reaction involves just one reactant.

33. Single and double displacement reactions are alike in that they always involve two reactants and two products.

34. When carbonates or bicarbonates react with acids, carbon monoxide is formed.

35. In the reaction $Cu(s) + 2\,AgNO_3(aq) \rightarrow Cu(NO_3)_2(aq) + 2\,Ag(s)$, Cu replaces Ag^+ because copper is a more reactive element than silver.

36. The combustion of magnesium is an endothermic reaction because magnesium must be heated to start the reaction.

37. The addition of sulfuric acid to water is an endothermic process.

38. The equation shown is balanced. $2\,Cu + 3\,HNO_3 \rightarrow Cu(NO_3)_2 + NO_2 + 2\,H_2O$.

39. The products for the complete combustion of butane (C_4H_{10}) are CO_2 and H_2O.

40. Decomposition reactions often require heat.

41. For a given chemical equation, there is only one possible set of whole numbers that can balance the equation.

42. When a piece of metallic magnesium is burned in air, it produces an intense fire and forms white magnesium oxide. This is an example of a single displacement reaction.

43. A cold pack applied to an injury is an example of an exothermic reaction.

44. When ammonia gas, NH_3, and oxygen gas, O_2, react to form nitrogen gas and water, for each mole of oxygen consumed, one mole of N_2 would be formed.

Multiple Choice (choose the best answer)

Identify each of the following unbalanced equations as being (a) a combination reaction, (b) a decomposition reaction, (c) a single displacement reaction, or (d) a double displacement reaction.

1. $Cl_2 + NaI \rightarrow NaCl + I_2$

2. $Ag_2O \rightarrow Ag + O_2$

3. $BaCl_2 + Na_2SO_4 \rightarrow BaSO_4 + NaCl$

4. $Na + Cl_2 \rightarrow NaCl$

5. $S + O_2 \rightarrow SO_2$

6. $H_2CO_3 \rightarrow H_2O + CO_2$

7. $Ba(ClO_3)_2 \rightarrow BaCl_2 + O_2$

8. $Zn + H_2SO_4 \rightarrow ZnSO_4 + H_2$

9. $SbCl_3 + Na_2S \rightarrow Sb_2S_3 + NaCl$

10. $C + O_2 \rightarrow CO_2$

Balance the equations below to answer the following question set (11–14):

> A. $Cl_2 + NaI \rightarrow NaCl + I_2$
> B. $Ag_2O \rightarrow Ag + O_2$
> C. $Ba(ClO_3) \rightarrow BaCl_2 + O_2$
> D. $SbCl_3 + Na_2S \rightarrow Sb_2S_3 + NaCl$

Add, for each equation, all coefficients (both reactants and products).

11.   The sum of the coefficients for reaction A is

>   a) 4   b) 6   c) 8   d) 5

12.   The sum of the coefficients for reaction B is

>   a) 3   b) 5   c) 7   d) 9

13. The sum of the coefficients for reaction C is

    a) 6   b) 4   c) 3   d) 5

14. The sum of the coefficients for reaction D is

    a) 12   b) 6   c) 9   d) 14

15. When the equation $K_2O + H_3PO_4 \rightarrow K_3PO_4 + H_2O$ is balanced, a term in the equation will be

    a) $3 H_3PO_4$ b) $2 K_3PO_4$ c) $2 K_2O$ d) $4 H_2O$

16. When the equation $BiCl_3 + H_2S \rightarrow Bi_2S_3 + HCl$ is the coefficient of HCl will be

    a) 3   b) 4   c) 5   d) 6

17. When the equation $Al + HCl \rightarrow$ (products) is completed and balanced, one term in the equation will be

    a) $AlCl_2$ b) $2 HCl$ c) $3 H_2$ d) $AlCl_3$

18. When the equation $KOH + H_2SO_4 \rightarrow$ (products) is completed and balanced, one term in the equation will be

    a) $2 H_2O$ b) KSO_4 c) H_2OH d) K_2SO_4

19. When the equation $Ca(OH)_2 + HBr \rightarrow$ (products) is completed and balanced, one term in the equation will be

    a) $CaBr_2$ b) $CaBr_3$ c) $3 HBr$ d) $H(OH)_2$

20. When a metal reacts with hydrochloric acid,

    a) a hydroxide is formed   b) a new acid is formed
    c) hydrogen is formed   d) a metal oxide is formed

21. When the following reaction is balanced with the lowest integer, the lowest coefficient of CO_2 is

    $C_4H_{10}(g) + O_2(g) \rightarrow CO_2(g) + H_2O(l)$

    a) 8   b) 16   c) 4   d) 10

22. When the following reaction is balanced with the lowest integer, the lowest coefficient of O_2 is

    $C_2H_5OH(g) + O_2(g) \rightarrow CO_2(g) + H_2O(l)$

    a) 1   b) 2   c) 3   d) 6

23. Incomplete combustion of methane produces carbon monoxide and water.
    $CH_4(g) + O_2(g) \rightarrow CO(g) + H_2O(l)$

    The coefficient of O_2 in the balanced equation is

    a) 4   b) 3   c) 2   d) 6

24. When solutions of an acid and a base are reacted, the products are

    a) a new acid and a new base   b) a salt and water
    c) a metal oxide and water   d) a nonmetal oxide and water

25. When Mg reacts with an aqueous solution of H_2SO_4, the products are

    a) $MgHSO_4 + H_2O$ b) $Mg(SO_4)_2 + H_2$
    c) $MgO + H_2 + SO_3$ d) No correct answer given.

26. $CO(g) + NO_2(g) \rightleftarrows CO_2(g) + NO(g) + 226$ kJ. The energy liberated when 2 moles of NO_2 react is

    a) 56.5 kJ   b) 113 kJ   c) 226 kJ   d) 452 kJ

27. Which of the following is not true about the reaction $N_2 + 2 O_2 + 68$ kJ $\rightarrow 2 NO_2$

    a) This is an exothermic reaction.
    b) 68 kJ of heat are needed for each mole of NO_2 formed.
    c) Two moles of O_2 react with one mole of N_2.
    d) Two moles of NO_2 are formed when one mole of N_2 reacts.

28. A combustion reaction always involves the production of

    a) sound   b) heat   c) oxygen gas   d) steam

29. Which of the following statements is not true during a chemical reaction?

    a) Total mass is conserved.
    b) The total number of atoms are conserved.
    c) The total number of molecules are conserved.
    d) Energy is involved.

30. The reaction of $NH_4Cl(aq)$ with $KOH(aq)$ produces

    a) $KCl(aq)$ and $NH_4OH(aq)$
    b) $KCl(aq)$ and $NH_3(g)$
    c) $KCl(aq)$, $NH_3(g)$ and $H_2O(l)$
    d) $KClO(aq)$, $NH_3(g)$ and $H_2(g)$

31. Consider the following reaction: $C(s) + O_2(g) \rightarrow CO_2 (g) + 393$ kJ
Which of the following statements is incorrect?

    a) The reaction is exothermic.
    b) The potential energy of the reactants is lower than the potential energy of the products.
    c) When 3 moles of C reacts with 3 moles of oxygen, 786 kJ of heat is released.
    d) This is a combustion reaction.

32. Consider the equation: $N_2 + H_2 \rightarrow NH_3$. Which of these statements is false?

    a) There is more than one set of coefficients which will balance this equation.
    b) This equation shows the decomposition of ammonia.
    c) This reaction involves both elements and compounds.
    d) When this equation is balanced, molecules will be conserved.

33. When the equation $Al + O_2 \rightarrow Al_2O_3$ is properly balanced, which of the following terms appears?

    a) $2\,Al$ b) $2\,Al_2O_3$ c) $3\,Al$ d) $2\,O_2$

34. Which equation is *incorrectly* balanced?

    a) $2\,KNO_3 \xrightarrow{\Delta} 2\,KNO_2 + O_2$
    b) $H_2O_2 \rightarrow H_2O + O_2$
    c) $2\,Na_2O_2 + 2\,H_2O \rightarrow 4\,NaOH + O_2$
    d) $2\,H_2O \xrightarrow[\text{H}_2\text{SO}_4]{\text{Electrical energy}} 2\,H_2 + O_2$

35. When the equation $F_2 + H_2O \rightarrow HF + O_2$ is balanced, a term in the balanced equation is

    a) $2\,HF$ b) $3\,O_2$ c) $4\,HF$ d) $4\,H_2O$

36. When the equation $H_3PO_4 + Ca(OH)_2 \rightarrow H_2O + Ca_3(PO_4)_2$ is balanced the proper sequence of coefficients is

    a) 3, 2, 1, 6   b) 2, 3, 6, 1   c) 2, 3, 1, 6   d) 2, 3, 3, 1

37. Choose the set of coefficients which balances the chemical equation given below.

$N_2O_5 + B_2O_3 + H_2O \rightarrow B_3N_3H_6 + O_2$

    a) 6, 5, 5, 2, 15   b) 3, 3, 3, 5, 2   c) 3, 3, 6, 2, 15   d) 2, 1, 1, 5, 1

38. Ferric oxide is reduced in the presence of carbon monoxide to produce metallic iron and carbon dioxide. This process is shown in the following unbalanced equation.

    ___ $Fe_2O_3 +$ ___ $CO \rightarrow$ ___ $Fe +$ ___CO_2

The number of molecules of CO required to form one atom of iron from its oxide is

    a) 1   b) 1.5   c) 2   d) 3

39. Methanol, CH_3OH, combusts in air to produce carbon dioxide and water vapor according to the following unbalanced equation:

    ___ $CH_3OH(l) +$ ___ $O_2(g) \rightarrow$ ___ $CO_2(g) +$ ___ $H_2O(g)$

If 6 moles of CH_3OH are burned in this way, how many moles of water will result?

    a) 4   b) 6   c) 8   d) 12

40. All of the following are true for carbon monoxide except

    a) CO results from incomplete combustion of methane
    b) CO is poisonous because it can bind to hemoglobin
    c) CO is colorless and has no smell
    d) CO is a laughing gas

41. All of the following are exothermic except

    a) freezing water
    b) burning wood
    c) dissolving salt in water
    d) reaction inside an ice pack

42. All of the following are endothermic except

    a) boiling water
    b) melting ice
    c) condensing water vapor
    d) reaction inside an ice pack

Completion

Write the symbol used in chemical equations that corresponds to the meaning of each of the following:

_____   **1.**   Yields, produces, (points to products)

_____   **2.**   Reversible reactions; equilibrium between reactants and products

_____   **3.**   Gas evolved

_____   **4.**   Solid or precipitate formed

_____   **5.**   Liquid (written after the substance)

_____   **6.**   Heat

_____   **7.**   Plus or added to

_____   **8.**   Aqueous solution (substance dissolved in water)

CHAPTER 9. Calculations From Chemical Equations

True — False (choose one)

1. Always write and balance the equation before you solve a stoichiometry problem.

2. The use of dimensional analysis is a good practice in solving stoichiometry problems.

3. In a balanced chemical equation the number of moles of products must be equal to the number of moles of reactants.

4. The molar mass of Na_3PO_4 is 163.9.

5. The molar mass of $Na_2CO_3 \cdot 10\ H_2O$ is 106.0.

6. A number in front of a formula in a balanced equation represents the number of moles of that substance in the chemical change.

7. A molecule is the smallest unit of a molecular substance and a mole is Avogadro's number of molecules of that substance.

8. The mass of a molecule of a substance is the same as the mass of a mole of the substance (e.g., 1 molecule of water is 18.02 g).

9. The mass of one molecule of water is $18.02\ g/6.022 \times 10^{23}$ molecules.

10. A mole of $Al(OH)_3$ contains the same number of oxygen atoms as a mole of Fe_2O_3.

11. A mole ratio is a ratio of the number of moles of reactants to the number of moles of products.

12. A mole ratio is a ratio between the number of moles of any two species involved in a chemical reaction.

13. In the reaction $C_3H_8 + 5\ O_2 \rightarrow 3\ CO_2 + 4\ H_2O$, the mole ratio for converting moles of O_2 to moles of H_2O is 5 mol O_2/4 mol H_2O.

14. In a chemical reaction at least one reactant must be in excess for the reaction to complete.

15. The limiting reactant is always in excess.

16. In a chemical reaction the limiting reactant determines how fast the reaction will occur.

17. It is possible under some circumstances for the actual yield of a chemical reaction to exceed the theoretical yield.

18. In a chemical reaction the amount of product that can be formed is determined by the limiting reactant.

19. When the yield of a chemical reaction is less than 100%, this indicates one of the reactants is present in insufficient amount to react with the entire amount of another reactant.

20. The reactant with the fewest number of moles is the limiting reactant in a chemical reaction.

21. In the reaction $N_2H_4 + 3 O_2 \rightarrow 2 NO_2 + 2 H_2O$, the limiting reactant is O_2 when 2 mol N_2H_4 and 3 mol O_2 are reacted.

22. In the reaction $2 Al + 3 Cl_2 \rightarrow 2 AlCl_3$, 0.50 mol of $AlCl_3$ was obtained when 1.0 mol of Cl_2 was reacted. The percent yield of $AlCl_3$ in the reaction is 75%.

23. Molecules are conserved in chemical reactions.

24. According to this reaction of the sugar sucrose, $C_{12}H_{22}O_{11} \rightarrow C + H_2O$, 12 moles of C result from the decomposition of one mole of sucrose.

25. To change from moles of the wanted in a chemical reaction to mass of the substance wanted, one must use the coefficients from the equation.

26. The symbol S_8 represents the same number of molecules in a chemical equation as 8 S.

27. In the reaction between aluminum and Fe_2O_3 to produce iron and aluminum oxide, 108 g of aluminum metal would completely react with 54 g of Fe_2O_3.

Multiple Choice (choose the best answer)

1. How many moles of $Na_2S_2O_3$ can be produced from 32.07 g of sulfur?

a) 0.500 mol   b) 1.00 mol   c) 32.07 mol   d) 158.1 mol

2. The number of grams of hydrogen in 0.80 mol of H_2 is

a) 1.6 g   b) 0.80 g   c) 3.2 g   d) 0.40 g

3. The number of moles in 12.0 g of C_2H_6O is

a) 12.0/24.02   b) 12.0 x 24.02   c) 12.0/46.07   d) 12.0 x 46.07

4. Which contains the larger number of molecules?

a) 16 g CH_4 b) 65 g SO_2 c) 40 g CO_2 d) 40 g CO

5. The number of moles in 1.00 kg NaCl is

a) 1000   b) 1000 x 58.44   c) 1.00/(58.44 x 1000)   d) 1000/58.44

6. The number of moles in 35.5 g of Cl_2 is the same as the number of moles in

a) 18.0 g H_2O b) 80.0 g $C_6H_{12}O_6$ c) 30.0 g $HC_2H_3O_2$ d) 40.1 g Ca

7. The mass of $Fe(OH)_3$ that can be produced from 10.0 g of $Fe_2(SO_4)_3$ is

a) 1.34 g   b) 2.67 g   c) 5.34 g   d) 2.04×10^3 g

8.   The number of hydrogen atoms in 25g of $C_6H_{12}O_6$ is

       a) 1.00×10^{24} b) 8.35×10^{22} c) 1.80×10^{26} d) 1.67

9.   How many moles of oxygen are needed to react with 1.6 moles of isopropyl alcohol, C_3H_7OH?
       Equation is $2 C_3H_7OH + 9 O_2 \rightarrow 6 CO_2 + 8 H_2O$

       a) 7.2   b) 0.36   c) 14.4   d) 4.5

10.   Given the equation: $Kr + 3 F_2 \rightarrow KrF_6$, how many moles of fluorine are required to produce 3.0 moles of krypton hexafluoride, KrF_6?

       a) 0.33   b) 1.0   c) 3.0   d) 9.0

11.   Given the unbalanced equation: $Ag_2CO_3 \rightarrow Ag + O_2 + CO_2$, how many moles of Ag can be produced from 2.5 moles of Ag_2CO_3?

       a) 2.50   b) 1.25   c) 5.00   d) 10.0

12.   How many moles of SO_2 can be produced from 20.0 g of H_2S?
              $2 H_2S + 3 O_2 \rightarrow 2 SO_2 + 2 H_2O$

       a) 0.312   b) 2.00   c) 20.0   d) 0.587

13.   How many moles of Fe_2O_3 will react with 85.2 g of Al?
              $2 Al + Fe_2O_3 \rightarrow Al_2O_3 + 2 Fe$

       a) 3.16   b) 1.58   c) 0.790   d) 0.500

14.   How many grams of O_2 are required to burn 25.0 moles of $C_{10}H_8$?
              $C_{10}H_8 + 12 O_2 \rightarrow 10 CO_2 + 4 H_2O$

       a) 9.60×10^3 b) 300   c) 66.7   d) No correct answer given.

15.   Consider the following reaction: $3 NO_2 + H_2O \rightarrow 2 HNO_3 + NO$
       What is volume of water needed to produce 2.00 kg of HNO_3? (The density of water = 1.00 g/mL)

       a) 572 mL   b) 28.6 mL   c) 286 mL   d) 35.9 mL

16.   How many grams of CO_2 can be produced from 25.0 moles of $C_{10}H_8$?
              $C_{10}H_8 + 12 O_2 \rightarrow 10 CO_2 + 4 H_2O$

       a) 250   b) 7.00×10^3 c) 1.10×10^4 d) 110

17.   How many grams of products can be produced from 3.55 moles of HgO?
              $2 HgO \rightarrow 2 Hg + O_2$

       a) 769   b) 1.54×10^3 c) 384   d) 826

18. How many grams of NaCN can be produced from 174 g of $Ca(CN)_2$?

$$Ca(CN)_2 + 2\,NaCl \rightarrow CaCl_2 + 2\,NaCN$$

    a) 185   b) 46.1   c) 2.2×10^3 d) 554

19. How many grams of O_2 are required to produce 100.g of SO_2?

$$4\,FeS_2 + 11\,O_2 \rightarrow 2\,Fe_2O_3 + 8\,SO_2$$

    a) 36.3   b) 49.9   c) 68.7   d) No correct answer given.

20. How many grams of ammonia can be produced when 20.0g of Mg_3N_2 and 20.0g of H_2O are reacted?

$$Mg_3N_2 + 6\,H_2O \rightarrow 3\,Mg(OH)_2 + 2\,NH_3$$

    a) 6.74   b) 3.37   c) 6.30   d) 18.9

21. When 75.0g NH_3, 90.0g CO_2, and 61.0g H_2O are reacted to form $(NH_4)_2CO_3$, the limiting reactant is _____. The balanced equation is $2\,NH_3 + CO_2 + H_2O \rightarrow (NH_4)_2CO_3$

    a) NH_3 b) CO_2 c) H_2O d) $(NH_4)_2CO_3$

Questions 22 – 25 pertain to the following balanced equation:

$$3\,Na_2S_2O_3 + 8\,KMnO_4 + H_2O \rightarrow 3\,Na_2SO_4 + 3\,K_2SO_4 + 8\,MnO_2 + 2\,KOH$$

22. The moles of Na_2SO_4 obtainable from the reaction of 3.0 moles of $Na_2S_2O_3$ with 3.0 moles of $KMnO_4$ and 3.0 moles of water is

    a) 3.0   b) 2.2   c) 1.1   d) 1.0

23. If the actual yield of Na_2SO_4 in the above question was 0.80 mole the percent yield is

    a) $\dfrac{3.0}{0.80} \times 100$ b) $\dfrac{0.80}{2.2} \times 100$ c) $\dfrac{0.80}{1.1} \times 100$ d) $\dfrac{0.80}{3.0} \times 100$

24. The reaction of 158 g of $KMnO_4$ will give

    a) 0.250 mol of KOH   b) 1.00 mol of KOH
    c) 0.500 mol of KOH   d) 2.00 mol of KOH

25. The total number of oxygen atoms that appear in the reactants of the equation is

    a) 40   b) 8   c) 42   d) 36

26. The limiting reactant is

    a) the reactant that yields the smaller amount of products
    b) the reactant that is completely consumed in the reaction
    c) the reactant that gives the theoretical yield
    d) a, b, c

27. Consider the following reaction $3A + B \rightarrow 3\,C$, where 2 moles of A react with 1 mole of B to completion. Which one of the following statements is correct?

   a) B is the limiting reactant because of its lower coefficient
   b) A is the limiting reactant because 2 moles of A will react with 1/3 mole of B
   c) B is the limiting reactant because you have fewer moles of B than A
   d) A is in excess

28. Consider the following reaction: $N_2(g) + 3\,H_2(g) \rightarrow 2\,NH_3\,(g)$

If 10.0 moles of nitrogen gas is mixed with 30.0 moles of hydrogen gas and the reaction goes to completion. Which of the following correctly describes the limiting reagent(s)?

   a) N_2 is the limiting reagent
   b) H_2 is the limiting reagent
   c) Both N_2 and H_2 are limiting reagents
   d) There is no limiting reactant because the reactants are present in stoichiometric ratio

29. One mole of SO_2 was obtained by the reaction of one mole of CS_2 with O_2 according to the equation

$$CS_2 + 3\,O_2 \rightarrow CO_2 + 2\,SO_2$$

   a) The yield of SO_2 is 100%
   b) The yield of SO_2 is 32.0g
   c) The yield of SO_2 is 64%
   d) More than 30g of oxygen reacted with the CS_2

30. Propane burns in excess oxygen according to the following reaction:

$$C_3H_8(g) + 5\,O_2(g) \rightarrow 3\,CO_2(g) + 4\,H_2O(l)$$

What amount of propane will produce 250g of CO_2 if the percent yield is 93.1%,?

   a) 89.7g   b) 269.1g   c) 244.8g   d) 78.9 g

31. Consider the following reaction: $2\,Mg + O_2 \rightarrow 2\,MgO$
If 4.0 moles of Mg react with 2.0 moles of O_2, which one of the following statements is correct?

   a) Both Mg and O_2 will be left over
   b) O_2 is the limiting reactant
   c) 2.0 moles of Mg will be left over
   d) 4.0 moles of MgO will be formed

32. $Ba(OH)_2 + Na_2SO_4 \rightarrow BaSO_4 + 2NaOH$

If 25g of $Ba(OH)_2$ is reacted, and the actual yield of $BaSO_4$ is 88%, the grams of $BaSO_4$ produced is

   a) 30 g   b) 34 g   c) 39 g   d) 22 g

33. $Fe_2(SO_4)_3 + 3\ Ba(OH)_2 \rightarrow 3\ BaSO_4 + 2\ Fe(OH)_3$

If 20 g of $Fe_2(SO_4)_3$ is mixed with 20 g of $Ba(OH)_2$, the limiting reactant is

 a) $Fe_2(SO_4)_3$ b) $Ba(OH)_2$ c) $Fe(OH)_3$ d) $BaSO_4$

34. $Mg(OH)_2 + 2HCl \rightarrow MgCl_2 + 2H_2O$

When 25.0 g $Mg(OH)_2$ and 25.0 g HCl are mixed and reacted, the amount of unreacted material will be

 a) 15.0 g $Mg(OH)_2$ b) 5.00 g $Mg(OH)_2$ c) 9.34 g HCl   d) 2.96 g HCl

35. $2\ Na_2O_2 + 2H_2O \rightarrow 4NaOH + O_2$

When 200. g of Na_2O_2 were reacted, 32.0 g of O_2 were collected. The yield of O_2 collected was

 a) 100%   b) 78.0%   c) 50.0%   d) No correct answer given.

36. For the reaction $C_2H_4 + O_2 \rightarrow CO_2 + H_2O$, suppose 18 moles of C_2H_4 reacted and you wanted to know how many moles of CO_2 would result. Which ratio should be used?

 a) 18 moles C_2H_4/1 mole CO_2 b) 1 mole C_2H_4/1 mole CO_2
 c) 2 moles CO_2/1 mole C_2H_4 d) 1 mole CO_2/2 moles C_2H_4

37. When $4Ag + 2H_2S + O_2 \rightarrow 2Ag_2S + 2H_2O$, and the reactants include 12 mol Ag, 6 mole of H_2S and 96g of O_2, the limiting reactant is

 a) Ag   b) H_2S c) O_2 d) No correct answer given.

The next three problems refer to the reaction $C_2H_4 + 3O_2 \rightarrow 2\ CO_2 + 2\ H_2O$

38. If 6.0 mol of CO_2 are produced, how many moles of O_2 were reacted?

 a) 4.0 mol   b) 7.5 mol   c) 9.0 mol   d) 15.0 mol

39. How many moles of O_2 are required for the complete reaction of 45 g of C_2H_4?

 a) 1.3×10^2 mol   b) 0.64 mol   c) 112.5 mol   d) 4.8 mol

40. If 14.0g of C_2H_4 is reacted and the actual yield of H_2O is 7.84g the percent yield in the reaction is

 a) 0.56%   b) 43.6%   c) 87.1%   d) 56.0%

41. $MgCl_2 + Na_2CO_3 \rightarrow MgCO_3 + 2\ NaCl$. When 10.0 g of $MgCl_2$ and 10.0 g of Na_2CO_3 are reacted, the limiting reactant is

 a) $MgCl_2$ b) Na_2CO_3 c) $MgCO_3$ d) NaCl

42. $Cu + 2\ AgNO_3 \rightarrow Cu(NO_3)_2 + 2\ Ag$. When 50.0 g of copper was reacted with silver nitrate solution, 148 g of silver was obtained. What is the percent yield of silver obtained?

 a) 87.1%   b) 84.9%   c) 55.2%   d) No correct answer given.

43. Sodium nitrate can be heated in an environment of excess hydrogen to form water. The two-step process is shown below:

$$2 \text{ NaNO}_3 \rightarrow 2\text{NaNO}_2 + \text{O}_2$$

$$2\text{H}_2 + \text{O}_2 \rightarrow 2\text{H}_2\text{O}$$

How many grams of sodium nitrate are required to form 90.0 grams of water?

a) 850   b) 690   c) 425   d) 213

44. Potassium chlorate undergoes a decomposition reaction to produce oxygen gas according to this chemical equation:

$$2 \text{ KClO}_3 \rightarrow 2 \text{ KCl} + 3 \text{ O}_2$$

If 2.8 kg of potassium chlorate are heated in this manner, the amount of oxygen that can be produced is

a) 364 g   b) 728 g   c) 1.09 kg   d) 2.18 kg

45. Phosphorus, P_4, burns in air to form diphosphorus pentoxide, P_2O_5. If 80.5 g of P_4, which is 90% pure, reacts, how many grams of oxygen would be required?

a) 18.7   b) 93.5   c) 104   d) 374

46. When iron (II) oxide, FeO, is reacted with hydrogen gas, the products are metallic iron and water. For each gram of FeO consumed, how many grams of iron are produced?

a) .0375 g   b) 0.50 g   c) 0.780 g   d) 1.0 g

CHAPTER 10. Modern Atomic Theory

True — False (choose one)

1. Wavelength is the distance a light travels between 2 consecutive peaks.

2. A photon is a particle-like bundle of electromagnetic radiation.

3. A ball rolling down a stairway is an example of a quantized change in energy.

4. The line spectrum of hydrogen is an evidence for the quantized energy for electrons.

5. In Bohr's atomic theory, energy is absorbed when an electron jumps from one principal energy level to a higher principal energy level.

6. The fourth energy level can contain a maximum of eighteen electrons.

7. The third energy level can contain 9 orbitals.

8. A 3p energy sublevel orbital can contain a maximum of six electrons.

9. An s orbital is spherically symmetrical around the nucleus.

10. An orbital is a region where the nucleus of an atom is likely to be found.

11. The maximum number of electrons that can occupy a specific energy level is given by the expression $2n^2$, where n is the principal energy level.

12. There are 11 protons in the nucleus of an atom having the electron configuration of $1s^2 2s^2 2p^6 3s^2 3p^4$.

13. The 4s energy sublevel is filled before electrons enter the 3d sublevel.

14. The 3d energy sublevel can have a maximum of 10 electrons.

15. The following series of energy sublevels is properly arranged in order of increasing energy 1s2s2p3s3p3d4s4p5s4d5p6s4f.

16. ⊞⊞⊞⊞⊞ represents the electron structure of chlorine.

17. No more than two electrons can occupy one orbital.

18. The 4d orbitals fill with electrons before the 5p orbitals.

19. The first element to have a d electron is scandium.

20. A 4d electron is in a lower energy level than a 3p electron.

21. All the noble gases have eight electrons in their outermost energy level.

22. The electron structure for a sulfur atom, $^{32}_{16}S$, is $1s^2 2s^2 2p^6 3s^2 3p^4$.

23. s and p orbitals are completely filled in the family of elements known as halogens.

24. Elements in the fourth period have the core electron configuration of argon.

25. Elements in a chemical family have similar atomic masses.

26. 3d orbitals are actually in an atom's fourth energy level.

27. Visible light represents the majority of all kinds of electromagnetic radiation known.

28. Hydrogen's single electron is most often found in the 1s orbital.

29. The period 3 metalloid is silicon.

30. The alkali metals are very chemically active because they have only one electron in their outermost shells.

31. The halogen elements all have one electron more than the nearest noble gas elements.

32. Selenium is an atom whose electrons occupy part or all of the fourth electron energy level.

33. A p orbital is spherically symmetrical around the nucleus.

34. The Lewis-dot symbol for potassium is $\cdot \ddot{\text{P}} \cdot$

35. A p type energy sublevel can contain six electrons.

36. A d type energy sublevel can contain five orbitals.

37. The maximum number of electrons in an f orbital is 14.

38. An atom with eight electrons in its outer shell is a noble gas.

39. The electron configuration of $_{22}$Ti is $1s^2 2s^2 2p^6 3s^2 3p^6 3d^4$.

40. An atom of an element of atomic number 22 has electrons in 11 orbitals.

41. The maximum number of electrons that can occupy a specific energy level $\mathbf{n}$ is given by $\mathbf{n}^2$.

42. Element Z, whose atoms have an outer-shell electron configuration of $ns^2 np^4$, is most likely to react chemically to form ions which have a charge of -2.

43. The ground state electron configuration of Zn is $1s^2 2s^2 2p^6 3s^2 3p^6 4s^2 3d^3$.

44. One would expect phosphorus to have chemical properties more similar to S than As.

45. Of the isoelectronic ions, Sc^{3+}, Ca^{2+}, K^+, and Cl^-, the one with the smallest ionic radius would be the scandium ion.

46. A hydrogen atom in its ground state can emit a photon.

Multiple Choice (choose the best answer)

1. All of the following are used to characterize electromagnetic radiation except

    a) wavelength   b) frequency   c) charge   b) speed

2. Which of the following is a correct atomic structure for $^{122}_{51}Sb$?

    a) $\left(\begin{array}{c}51p\\51n\end{array}\right)$ 2 8 18 18 5   b) $\left(\begin{array}{c}51p\\122n\end{array}\right)$ 2 8 18 18 5

    c) $\left(\begin{array}{c}51p\\71n\end{array}\right)$ 2 8 18 18 5   d) $\left(\begin{array}{c}51p\\51n\end{array}\right)$ 2 8 18 8 5

3. The number of electrons a single d orbital can hold is

    a) 10   b) 6   c) 2   d) 14

4. How many d electrons can the second energy level hold?

    a) 0   b) 2   c) 6   d) 10

5. The number of s electrons in a single oxygen atom is

    a) 0   b) 2   c) 4   d) 6

6. What is the total number of electron containing orbitals in a nitrogen atom?

    a) 5   b) 3   c) 4   d) 6

7. Of the orbitals shown, the one with the lowest energy level is

    a) 2s   b) 3s   c) 3d   d) 3p

8. The maximum number of electrons that can occupy a 3p orbital is

    a) 1   b) 2   c) 3   d) 6

9. The electron configuration of an atom is $1s^2 2s^2 2p^6 3s^2 3p^3$. The atomic number of the atom is

    a) 15   b) 11   c) 5   d) 3

10. The electron configuration of an atom is $1s^2 2s^2 2p^6 3s^2 3p^6$. The number of unpaired electrons in this atom are

    a) 2   b) 3   c) 5   d) No correct answer given.

11. The electron configuration of an atom is $1s^2 2s^2 2p^6 3s^2 3p^6$. The number of orbitals occupying electrons is

    a) 5   b) 9   c) 11   d) 15

12. The maximum number of electrons in the 4d sublevel is

    a) 2   b) 6   c) 8   d) 10

13. Which is the electron configuration for an atom of atomic number 18?

    a) $1s^2 2s^2 2p^6 3s^2 3p^2$ b) $1s^2 2s^2 2p^6 3s^2 3p^4 4s^2$

    c) $1s^2 2s^2 2p^6 3s^2 3p^6$ d) $1s^2 2s^2 2p^6 3s^2 3p^6 4s^2$

14. The correct electron sublevel structure for $_{25}$Mn is

    a) $1s^2 2s^2 2p^6 3s^6 3d^9$ b) $1s^2 2s^2 2p^6 2d^{10} 3s^2 3p^3$

    c) $1s^2 2s^2 2p^6 3s^2 3p^6 4s^2 4p^5$ d) $1s^2 2s^2 2p^6 3s^2 3p^6 4s^2 3d^5$

15. $[Ar]4s^2 3d^2$ is the electron configuration for

    a) Scandium   b) Titanium   c) Vanadium   d) Zirconium

16. The correct electron configuration for chromium is

    a) $[Ar] 4s^2 3d^4$ b) $[Ar] 4s^2 4d^4$ c) $[Ar] 4s^1 3d^5$ d) $[Ar] 3d^6$

17. Which of the following orbital diagrams is not possible for an atom in a ground state?

    a) $\uparrow\downarrow$ $\uparrow\downarrow$ $\uparrow\downarrow$ $\uparrow\downarrow$ $\uparrow\downarrow$
         1s   2s   2p

    b) $\uparrow\downarrow$ $\uparrow\downarrow$ $\uparrow$ $\uparrow$ $\uparrow$
         1s   2s   2p

    c) $\uparrow\downarrow$ $\uparrow$ $\uparrow\downarrow$ $\uparrow\downarrow$ $\uparrow\downarrow$
         1s   2s   2p

    d) $\uparrow\downarrow$ $\uparrow\downarrow$ $\uparrow\downarrow$ $\uparrow$ $\uparrow$
         1s   2s   2p

18. Which of the following orbital diagrams is not possible for an atom in a ground state?

    a) $\uparrow\downarrow$ $\uparrow\downarrow$ $\uparrow\downarrow$ $\uparrow\downarrow$ $\uparrow\downarrow$
         1s   2s   2p

    b) $\uparrow\downarrow$ $\uparrow\downarrow$ $\uparrow$ $\uparrow$ $\uparrow$
         1s   2s   2p

    c) $\uparrow\downarrow$ $\uparrow\downarrow$ $\uparrow\downarrow$ $\uparrow\downarrow$ $\uparrow\downarrow$
         1s   2s   2p

    d) $\uparrow\downarrow$ $\uparrow\downarrow$ $\uparrow\uparrow$ $\uparrow$ $\uparrow$
         1s   2s   2p

19. Which one of the following orbitals represents the lowest excited state of hydrogen?

    a) $1s^1$ b) $2s^1$ c) $2p^1$ d) $3s^1$

20. Which one of the following orbitals represents the highest excited state of hydrogen?

    a) $1s^1$ b) $2s^1$ c) $2p^1$ d) $3s^1$

21. $[Kr]5s^2 4d^{10}$ is the electron configuration for

    a) Br   b) I   c) At   d) Cd

22. Which atom does not have an unpaired electron in its ground-state electron structure?

    a) C   b) Ca   c) P   d) Cl

23. As atomic number increases across a period, all of the following increase *except*

    a) atomic radius   b) atomic mass   c) ionization energy   d) number of valence electrons

24. As atomic number increases down a family, all of the following increase *except*

    a) atomic radius   b) atomic
    c) number of valence electrons   d) No correct answer given.

25. Which of the waves is not a part of the electromagnetic spectrum?

    a) x-rays   b) microwaves   c) sound waves   d) red light

26. Regions of space around the nucleus of an atom that can be occupied by one or two electrons with identical energy are called

    a) shells   b) energy levels   c) photons   d) orbitals

27. The first energy level which has f orbitals to fill is

    a) 2   b) 3   c) 4   d) 5

28. Members of oxygen's family on the periodic table have electron configurations which end with

    a) ns^2 b) $ns^2 np^1$ c) $ns^2 np^2$ d) $ns^2 np^4$

29. An atom in the excited state might have the configuration

    a) $1s^2 2s^3$ b) $1s^2 2s^2 3d^1$ c) $1s^1 2s^2 2p^2$ d) $1s^2 2s^2 2p^1$

30. The total number of orbitals that contain at least one electron in an atom having the structure $1s^2 2s^2 2p^6 3s^2 3p^2$ is

    a) 5   b) 8   c) 14   d) No correct answer given.

31. Which is not a correct Lewis-dot symbol for phosphorus, $_{15}P$?

    a) :P̣·   b) :P̈·   c) ·P̈·   d) ·Ṗ:

32. Which of the following is the correct Lewis-dot symbol for bromine?

    a) Br·   b) Ḅr·   c) :Ḅr·   d) :B̈r:

33. How many unpaired electrons are in the electron structure of $_{24}Cr$, [Ar] $4s^1 3d^5$?

    a) 2   b) 4   c) 5   d) 6

34. Which of the following sets shows three isoelectronic species?

    a) Cu, Ag, Au   b) H^-, Li^+, Be^{+2} c) Cl^+, Ar, K^- d) P^{-3}, Se^{-2}, I^-

35. Proceeding from left to right across the Periodic Table, which of these statements is true?

    a) The atomic size of the elements increases.
    b) The electronegativity of the elements decreases.
    c) The valence electrons of the elements are held more weakly.
    d) The ionization energy of the elements increases.

CHAPTER 11.   Chemical Bonds: The Formation of Compounds from Atoms

True — False (choose one)

1. The elements of Group VA of the periodic table have five valence electrons.

2. Metals generally gain electrons when they react and become positive ions.

3. Cations are negative ions.

4. The anion of an atom of an element is larger than a neutral atom of the element.

5. Reactions between elements differing greatly in electronegativities result in ionic bonding between the elements reacting.

6. Transfer of electrons between atoms results in covalent bonding between the atoms.

7. Ionically bonded substances do not exist as molecules.

8. A sulfide ion (S^{2-}) has less electrons than a sulfur atom.

9. A calcium ion (Ca^{2+}) has less electrons than a calcium atom.

10. The Cl^-, S^{2-}, and Na^+ ions have the same electron structure.

11. A sodium atom is larger than a chlorine atom but a sodium ion is smaller than a chloride ion.

12. A magnesium atom loses two electrons when it reacts with chlorine to form $MgCl_2$.

13. Based on the periodic table, the formula for calcium bromide would be $CaBr_2$.

14. Based on the periodic table, the formula for calcium nitride would be Ca_3N_2.

15. NaO_3 is the correct formula for the compound formed in the reaction between sodium and oxygen.

16. If the formula for sulfuric acid is H_2SO_4, then the formula for selenic acid will likely be H_2SeO_4.

17. The bond between the two chlorine atoms in Cl_2 is covalent.

18. A covalent bond is the result of the electrostatic attraction between oppositely charged ions.

19. A pair of electrons shared between two atoms constitutes a covalent bond.

20. Large aggregates of positive and negative ions make up molecules.

21. All these compounds contain covalent bonds: H_2, HCl, CH_4, SO_2, N_2O_3.

22. When the elements boron and fluorine react and combine, the expected product is BF_3.

23.   In most covalent bonds, the pairs of electrons are shared equally.

24.   The electronegativity of an atom is a relative value representing the attraction that atom has for shared electrons in a molecule.

25.   Metallic elements tend to have high electronegativity values.

26.   The highest negativity on the Pauling scale is 4.0.

27.   The three most electronegative elements are O, F, Cl.

28.   The Lewis structure for HBr is H $:\overset{..}{\underset{..}{Br}}:$

29.   A nitrogen molecule, N_2, has three pairs of shared electrons and two pairs of unshared electrons.

30.   The Lewis structure for a phosphate ion, PO_4^{-3}, is $\begin{smallmatrix} :\overset{..}{O}: \\ :\overset{..}{O}:\overset{..}{P}:\overset{..}{O}:^{3-} \\ :\overset{..}{O}: \end{smallmatrix}$

31.   The Lewis structure for hydrazine, N_2H_4, is $\begin{smallmatrix} H:\overset{..}{N}:\overset{..}{N}:H \\ \overset{..}{H}\ \overset{..}{H} \end{smallmatrix}$

32.   Unequal sharing of the electron pairs in a covalent bond causes the bond to be nonpolar.

33.   All molecules containing polar bonds will be polar molecules.

34.   Polar molecules contain polar bonds.

35.   The chemical bonds in a water molecule are polar covalent.

36.   A carbon monoxide molecule, CO, is a dipole.

37.   An H_2S molecule is polar.

38.   All these compounds contain polar covalent bonds: H_2, HCl, CH_4, SO_2, N_2O_3.

39.   All polyatomic ions are positively charged.

40.   A molecule with a central atom surrounded by three pairs of electrons will have a trigonal planar shape.

41.   A molecule with a central atom surrounded by two bonding and two nonbonding pairs of electrons will have a tetrahedral shape.

42.   A molecule with a central atom surrounded by three bonding and 1 nonbonding pairs of electrons will have a tetrahedral electron structure.

43. H_2S and CO_2 both have a bent shape.

44. The type of bond that results from an equal sharing of electrons is polar.

45. A molecule shows resonance if it has multiple correct Lewis structures.

46. CO_3^{2-} has three resonance structures.

47. The Lewis structure for CO_2 is $:\ddot{O}:C:\ddot{O}:$

48. When a central atom in a molecule is connected to two other atoms and also has two pairs of non-bonded electrons, that molecule is most likely bent.

49. The electrons in the outermost shell of an element are called the valence electrons.

50. When an atom loses an electron it becomes a negative ion.

51. Crystals of sodium chloride consist of arrays of NaCl molecules.

52. The sharing of a pair of electrons between a positive ion and a negative ion is called an ionic bond.

53. Bromine has a greater electronegativity than chlorine.

54. A dipole is a molecule that is electrically asymmetrical, causing it to be oppositely charged at two points.

55. The water molecule is a dipole.

56. A stable group of atoms that has either a positive or a negative charge and behaves as a single unit is called a polyatomic ion.

57. The arrangement of electron pairs around an atom determines its shape.

58. Nonbonding electron pairs are ignored when determining the shape of a molecule.

59. A sodium ion is smaller than a sodium atom.

60. In crystals of sodium chloride, each sodium ion is surrounded by six chloride ions, and each chloride ion is surrounded by six sodium ions.

61. Boron has a greater electronegativity than barium.

62. When metals with low electronegativity combine with nonmetals of high electronegativity, they tend to form ionic compounds.

63. A pair of electrons shared between two atoms constitutes a covalent bond.

64. The covalent bond between a B atom and an N atom will be nonpolar.

65. The most stable arrangement of the atoms in a molecule of phosphorus trichloride would be square planar.

66. The ability that any atom has to compete for electrons with other atoms to which it is bonded is best described as its ionization energy.

67. All three of the molecules: H_2S, HCN, and CO are linear in shape.

Multiple Choice (choose the best answer)

1. In becoming a cation, an atom

    a) loses electrons   b) gains electrons
    c) shares electrons   d) No correct answer given.

2. In becoming an anion, an atom

    a) loses electrons   b) gains electrons
    c) shares electrons   d) No correct answer given.

3. Which atom has four valence electrons?

    a) Be   b) Fe   c) Ge   d) Te

4. Which atom does not have the same number of valence electrons as the other atoms given?

    a) O   b) N   c) S   d) Se

5. Which has the largest radius?

    a) Cl^- b) Cl   c) Na^+ d) Na

6. Which pair of elements exhibits the highest size difference?

    a) H and Rb   b) He and Rb   c) He and Fr   d) cannot tell

7. In which pair does the ion have a larger radius than the atom?

    a) Na and Na^+ b) Br and Br^- c) Al and Al^{3+} d) H and H^+

8. When a magnesium atom changes to a magnesium ion, its radius

    a) increases   b) decreases   c) remains the same   d) insufficient data to answer

9. When an atom loses two electrons, it becomes an ion with a charge of

    a) 0   b) +2   c) - 2   d) +4

10. During the formation of an ionic bond, the atom which transfers its valence electron(s) to another atom is the atom with the

    a) higher electronegativity   b) lower electronegativity
    c) higher ionization energy   d) No correct answer given.

11. The metal $^{45}_{21}Sc$ loses three electrons to form an ion. The ion has the formula

    a) Sc^{3+} b) Sc^{3-} c) Sc^{4-} d) Sc^{4+}

12. Which of the following does not have a noble gas electron structure?

   a) S^{2-} b) Ar   c) Al^{3+} d) Sb^{5+}

13. Which elements react with the halogens forming compounds containing a ratio of 1 halogen atom to 1 atom of the other element?

   a) alkali metals   b) alkaline earth metals   c) metalloids   d) noble gases

14. The formula of the compound formed in the reaction between lithium and sulfur is

   a) LiS   b) LiS_2 c) Li_2S_3 d) Li_2S

15. The formula of the compound formed in the reaction between barium and oxygen is

   a) BaO   b) Ba_2O c) BaO_2 d) Ba_2O_3

16. Atoms in a stable compound achieve the nearest noble gas configuration by

   a) sharing electrons   b) donating electrons   c) accepting electrons   d) a, b, c.

17. Metal and non-metal react to form

   a) molecular compound
   b) ionic compound
   c) hydrate compound
   d) molecular and ionic compound

18. Sodium and chlorine react to form sodium chloride. Which of the following statements is correct?

   a) The sodium ion has the electronic configuration of Ne.
   b) The chlorine ion has the electronic configuration of Ar.
   c) The sodium ion is smaller than the sodium atom.
   d) All are correct.

19. Which formula does *not* represent a correct compound of magnesium?

   a) $MgCl_2$ b) Mg_2S c) MgO   d) Mg_3N_2

20. Which formula is incorrect?

   a) RbFr   b) RbAt   c) CaI_2 d) FrAt

21. A covalent bond must involve

   a) two protons   b) two electrons   c) two neutrons   d) four protons

22. Which formula represents a true molecule?

   a) NaCl   b) Na_2 c) F_2 d) KF

23. To break or split a covalent bond, energy must be

a) absorbed   b) released   c) both absorbed and released   d) created

24. The atoms in an I_2 molecule are held together by a(n)

a) ionic bond   b) coordinate covalent bond   c) covalent bond   d) halogen bond

25. The bond between the atoms in HCl is a(n)

a) polar covalent bond   b) nonpolar covalent bond   c) ionic bond   d) double bond

26. Elements that have low electronegativity have

X. low electron affinity.   Y. low ionization energy.   Z. bigger radius.

a) X, Y   b) Y, Z   c) X, Y, Z   d) X, Z

27. Elements that have low ionization energy have

X. high electron affinity.   Y. high electronegativity.   Z. bigger radius.

a) X, Y   b) Z   c) X, Y, Z   d) X, Z

28. Which of the following is the most electronegative?

a) Na   b) Rb   c) Cl   d) Se

29. Which of the following is the most electronegative?

a) C   b) Rb   c) Cl   d) Se

30. Which of the following is the least electronegative?

a) N   b) P   c) As   d) Sb

31. Which pair of elements will form the most ionic bond?

a) C and Cl   b) N and O   c) Na and I   d) Al and Br

32. If the electronegativity difference between two bonding atoms is sufficiently large

a) a nonpolar covalent bond forms   b) a coordinate covalent bond will result
c) an ionic bond forms   d) a double bond will result

33. The number of valence electrons in the atoms of HNO_3 is

a) 18   b) 20   c) 23   d) 24

34. The number of valence electrons in a bromate ion, BrO_3^-, is

a) 23   b) 24   c) 25   d) 26

35. The Lewis structure for an ammonium ion, NH_4^+, is

a) H:N̈:H with H below, +   b) H:N̈:H with H below, +   c) H:N̈:H with H below, +   d) H·Ḧ·H with H below, +

36. Which Lewis structure is incorrect?

a) Cl^-, :C̈l:⁻   b) S^{2-}, :S̈:²⁻   c) HCl, H:C̈l: d) N^{3-}, :N̈:³⁻

37. The Lewis structure for the nitrite ion, NO_2^-, is

a) :Ö:N̈:Ö:⁻   b) :Ö::N̈::Ö:⁻   c) :Ö::N̈:Ö:⁻   d) :Ö:N̈:Ö:⁻

38. A molecule of hypochlorous acid consists of one atom each of hydrogen, chlorine, and oxygen. The most likely Lewis structure for this acid is

a) H:C̈l:Ö: b) :C̈l:H:Ö: c) :C̈l:Ö:H   d) C̈l:Ö:H:

39. Experiment shows that H_2O is a dipole and CO_2 is not a dipole. The two structures which best illustrate these properties are

a) $O = C = O$, $H - O - H$ b) H⌐O⌐H, O⌐C⌐O

c) $H - H - O$, $O = C = O$ d) $O = C = O$, H⌐O⌐H

40. The fact that carbon disulfide, CS_2, is not a dipole indicates that the molecule is

a) angular   b) covalently bonded   c) ionically bonded   d) linear symmetrical

41. Which of the following contains a nonpolar covalent bond?

a) HCl b) Cl_2 c) Cl_2O d) NH_3

42. Which of the following is a polyatomic ion?

a) Al^{3+} b) N_2 c) SO_3 d) NO_3^-

43. Which of the following molecules shows three resonances?

I. NO_3^- II. CO_3^{2-} III. CO_2

a) I, III   b) I, II   c) II   d) I, II, III

44.   Which of the following molecules/ions is/are linear?
        I. NO_2^- II. $SOCl_2$ III. Cl_2O

        a) I   b) I and II   c) II and III   d) III

45.   Which of the following compound is/are polar?
        I. H_2O II. CO_2 III. CF_4

        a) III   b) II and III   c) II   d) I

46.   In the reaction between barium atoms and sulfur atoms the barium atoms

        a) become covalent
        b) become part of polyatomic ions
        c) become cations
        d) share electrons with sulfur

47.   Which compound has a trigonal planar shape?

        a) H_2O b) $BeCl_2$ c) AlF_3 d) NH_3

48.   The bond angle between fluorine atoms in CF_4 is about

        a) 90°   b) 105°   c) 109°   d) 180°

49.   Which of the following formulas is not correct?

        a) Na^+ b) S^- c) Al^{3+} d) F^-

50.   Which of the following does not have a polar covalent bond?

        a) CH_4 b) H_2O c) CH_3OH d) Cl_2

51.   Which of the following has bonding that is ionic?

        a) H_2 b) MgF_2 c) H_2O d) CH_4

52.   Which of the following is an incorrect Lewis structure?

        a) H:N̈:H   b) :Ö:H   c) H:C̈:H   d) :N:::N:
            Ḧ   Ḧ   Ḧ

53.   Which element has seven valence electrons?

        a) S   b) Ne   c) Br   d) Ag

54.   Which of the following would be an incorrect formula?

        a) NaCl   b) K_2O c) AlO   d) BaO

55. As the difference in electronegativity between two elements decreases, the tendency for the elements to form a covalent bond

    a) increases   c) remains the same
    b) decreases   d) sometimes increases and sometimes decreases

56. Which compound has a bent (V-shaped) molecular structure?

    a) $NaCl$ b) CO_2 c) CH_4 d) H_2O

57. The total number of valence electrons in a nitrate ion, NO_3^-, is

    a) 12   b) 18   c) 23   d) 24

58. The number of lone (unbonded) pairs of electrons in H_2O is

    a) 0   b) 1   c) 2   d) 4

59. Which of these fluoride compounds would be expected to have a planar shape?

    a) SF_6 b) CF_4 c) BF_3 d) NF_3

60. HNO_2 is known as nitrous acid; the most reasonable Lewis structure for this molecule would be

    a) H:Ö:::N:Ö: b) H:Ö::N::Ö: c) H:Ö:Ṅ::Ö: d) H:Ö:Ṅ:Ö:

61. Which of these compounds would have a linear shape?

    a) O_3 b) SO_2 c) H_2O d) CO_2

62. Dihydrogen trioxide, H_2O_3, is a highly unstable and volatile compound which is related to hydrogen peroxide. Which of the following Lewis structures would be most reasonable for its molecular structure?

    a) H:Ö:Ö:Ö:H   b) :Ö:H:Ö:H:Ö: c) :H::O::O::O::H: d) :O::H::O::H::O:

63. Nitrogen trifluoride, NF_3, and boron trifluoride, BF_3, are similar in composition, but their molecular shapes are different. NF_3 forms a pyramid while BF_3 is trigonal planar. Which of these statements explains this difference?

    a) N is more electronegative than B.
    b) B uses three valence orbitals in bonding and N does not.
    c) BF_3 is ionic; NF_3 is covalent.
    d) N has a nonbonding pair of valence electrons, while B does not.

Matching

From the list of terms given, choose the one that correctly identifies each phrase.

List of terms: (a) polar covalent bond; (b) electron affinity; (c) nonpolar covalent bond; (d) valence electrons; (e) polyatomic ion; (f) covalent bond; (g) dipole; (h) electronegativity; (i) ionic bond; (j) Lewis structure; (k) trigonal planar; (l) tetrahedral.

_____ 1.   The attractive force that an atom of an element has for shared electrons in a molecule.

_____ 2.   Electrons in the outermost shell of an atom.

_____ 3.   The type of covalent bond formed between two atoms of different electronegativities.

_____ 4.   Bond formed when a complete transfer of an electron (or electrons) takes place from one atom to another.

_____ 5.   A stable group of atoms that behaves as a single unit and has either a positive or a negative charge.

_____ 6.   The shape of the CBr_4 molecule.

_____ 7.   The type of bond formed by the sharing of electrons between atoms.

_____ 8.   A molecule that is electrically asymmetrical, causing it to be oppositely charged at two points.

CHAPTER 12.   The Gaseous State of Matter

True — False (choose one)

1.   Of the three states of matter, gases are the least compact and the most mobile.

2.   Gases are capable of being greatly compressed.

3.   Molecules of O_2 gas and H_2 gas at the same temperature will have the same average kinetic energies and the same average velocities.

4.   In a large container of O_2 gas the pressure exerted by the oxygen will be greater at the bottom of the container.

5.   All collisions between gas molecules are perfectly elastic (no energy lost) according to the Kinetic Molecular Theory.

6.   The average kinetic energy of molecules is directly proportional to the Kelvin temperature.

7.   The assumption of the Kinetic Molecular Theory that gas molecules have no attraction for each other is not always valid.

8.   The property of diffusion is the ability of two or more gases to spontaneously mix.

9.   The rates of effusion of two different gases are directly proportional to their densities.

10.   Pressure is defined as force per unit volume.

11.   The kinetic energy of a gas depends on the number of gas molecules present.

12.   Gases can be compressed because their molecules are very tiny.

13.   One assumption of the kinetic molecular theory is that gas molecules do not lose kinetic energy when they collide with one another.

14.   A real gas most closely represents an ideal gas at high temperature and low pressure.

15.   The fact that a balloon filled with helium gas will leak more slowly than one filled with hydrogen gas is explained by Graham's Law of Effusion.

16.   At the same temperature and pressure methane (CH_4) will effuse faster than ethane (C_2H_6).

17.   The rate of effusion of helium (He) is about four times that of sulfur dioxide (SO_2).

18.   The pressure of a gas can be measured with a gauge called a manometer, or a barometer.

19.   Atmospheric pressure varies with altitude.

20.   When the temperature of a gas in a fixed volume is increased, the pressure of the gas increases.

21. Given one mole of a gas in a one liter container. The pressure of the gas can be increased by adding more gas molecules or decreasing the temperature.

22. The pressure exerted by a gas is directly proportional to the number of gas molecules present.

23. The two most abundant gases in the atmosphere are oxygen and carbon dioxide.

24. The following pressures are equal to one another: 1 atm; 1 toff; 760 mm Hg; and 29.9 in. Hg.

25. One mole of hydrogen gas in a one liter container exerts the same pressure as one-half mole hydrogen gas in a two liter vessel.

26. The volume of a gas is inversely proportional to the number of moles of gas present (pressure and temperature constant).

27. A mathematical statement of Boyle's Law is $\frac{P}{V} = k$.

28. The volume of gas at constant pressure is inversely proportional to the absolute temperature of the gas.

29. At absolute zero the volume of an ideal gas would become zero.

30. The French physicist J. A. C. Charles found that various gases expanded by the same fractional amount when heated through the same temperature interval.

31. Mathematically, Charles's Law states $V = kT$ (P constant).

32. If the Celsius temperature is doubled, the pressure of a fixed volume of gas would increase by the ratio of the temperatures in Kelvins.

33. When you triple the pressure on a gas in a variable volume container, the volume will decrease to one-third of the original volume.

34. Standard temperature and pressure (STP) are 0°C and 760 atm pressure.

35. Standard temperature and pressure are 0°C and 1 atm pressure.

36. A mixture of one mole of H_2 and one mole of CH_4 at STP will occupy a volume of 44.8 L.

37. A mixture of 0.5 mol O_2 and 0.5 mol H_2 in a one liter container at 0°C will exert a pressure of 22.4 atm.

38. The total pressure of a mixture of gases is the sum of the partial pressures of all the gases present in the container.

39. When measured at constant temperature and pressure, the ratios of the volumes of reacting gases are small whole numbers.

40. Equal volumes of the same gas at the same temperature and pressure contain the same number of molecules, the same number of moles, and the same mass.

Chapter 12. The Gaseous State of Matter

41. Equal volumes of NO_2 and CO_2 gases at the same temperature and pressure contain the same number of moles of gases.

42. According to Avogadro's Law, there are 6.022×10^{23} molecules of gas in a 22.4 L container at STP.

43. 2.24 L of O_2 at 0°C and 2 atm pressure contain 0.100 mole of O_2.

44. The volume of one mole of an ideal gas is always 22.4 L.

45. To produce 10 L of NH_3 from N_2 and H_2, [$N_2(g) + 3H_2(g) \rightarrow 2NH_3(g)$], would require 15 L of H_2 gas.

46. The molar volume of any gas is 22.4 L at STP.

47. 44.0 g of CO_2 occupy a volume of 22.4 L at STP.

48. The densities of gases are usually expressed in g/mL.

49. The density of nitrogen gas at STP is 28.0 g/22.4 L.

50. The average density of air is 1.29 g/L.

51. The gases CO_2, C_3H_8, and N_2O all have the same density.

52. In describing the behavior of gases, the ideal gas equation relates the variables of temperature, pressure, volume and moles.

53. The volume of a gas depends only on the temperature and pressure.

54. The gas law constant, R, can have different values, depending on the units of temperature and pressure used.

55. The volume of one mole of a gas can be calculated from $V = \dfrac{nRT}{P}$.

56. For reacting gases: volume-volume relationships are the same as the mole-mole relationships.

57. The formation of O_3 from O_2 is an exothermic reaction.

58. Infrared radiation from the sun converts oxygen to ozone in the stratosphere.

59. Chlorofluorocarbons that get to the stratosphere react to destroy the protective ozone layer.

60. Ozone in the stratosphere regulates the amount of ultraviolet radiation that reaches the earth.

61. At any given temperature and pressure, all gas molecules in a mixture will have the same average speed.

62. Gas molecules exert pressure when they collide with the walls of the container they are in.

63. The density of any gas is directly related to its molar mass at any given temperature and pressure.

64. An appropriate unit for pressure would be pounds/feet.

65. The average kinetic energy of molecules of a gas increases with increased temperature.

66. One mole of hydrogen gas and 1 mole of oxygen gas, each in a box of equal volume, and each at the same temperature, will exert the same pressure.

67. One mole of any gas always occupies 22.4 liters.

68. When the temperature of a gas is decreased at fixed volume, the pressure decreases.

69. A barometer is an instrument used to measure the mass of a certain quantity of mercury.

70. Chlorofluorocarbons, which were used in aerosol spray cans and are used in refrigeration and air conditioners, escape to the stratosphere and react to destroy part of the ozone layer.

71. The formula of ozone is 3 O_2.

72. The phenomenon in which an element can exist in two or more molecular or crystalline forms is known as allotropy.

73. The density of a sample of an unknown gas is measured at 65.0°C and 730 mm Hg and is found to be 2.14 g/L. Under the very same conditions, the density of oxygen gas is 1.11 g/L. The molar mass of the unknown gas must be 16.6 g/mol.

74. There are 2.69 x 10^{19} molecules in 1.00 mL of an ideal gas at STP.

75. Under certain conditions real gases exhibit measured pressures less than those predicted by the Ideal Gas Law because intermolecular attractions exist between real molecules.

76. Three liters of sulfur dioxide gas, SO_2, and two liters of oxygen gas, O_2, are reacted at a certain temperature and pressure. Under the same conditions, one can expect five liters of sulfur trioxide gas, SO_3, to form.

77. Exactly 500 mL of a gaseous compound has a mass of 0.9825 at STP. The molar mass of the gas is approximately 44.0 g/mol.

78. A gas which will diffuse twice as fast as SO_2 is O_2.

79. The density of a gas is directly related to its molar mass, to the pressure, and to the temperature.

80. Ideal gas molecules are held together by the same type of intermolecular forces that exist in liquid.

81. Real gas molecules are held together by the same type of intermolecular forces that exist in liquid.

Multiple Choice (choose the best answer)

1. The conditions under which gases behave most ideally are

   a) low temperature, low pressure   b) high temperature, high pressure
   c) high temperature, low pressure   d) low temperature, high pressure

2. The conditions under which gases behave least ideally are

   a) low temperature, low pressure   b) high temperature, high pressure
   c) high temperature, low pressure   d) low temperature, high pressure

3. Which of the following is incorrect regarding atmospheric pressure?

   a) Atmospheric pressure is always equal to 1 atm.
   b) Atmospheric pressure varies with elevation.
   c) Atmospheric pressure is equal to 1 atm at sea level.
   d) Atmospheric pressure is close to 0 atm at an altitude of 27 miles.

4. Which of the following defines standard conditions for a gas?

   a) 0 K and 1 atm   b) $273^{\circ}C$ and 760 torr   c) 273 K and 760 torr   d) $25^{\circ}C$ and 1 atm

5. If the rate of effusion of methane gas (CH_4) is about two times that of gas X, the molar mass of gas X is

   a) 32   b) 4   c) 64   d) 16

6. Which pressure does not correspond to 1.25 atm?

   a) 950 torr   b) 37.4 in. Hg   c) 950 cm Hg   d) 1.27 kPa

7. Given a 500. mL sample of H_2 at 2.00 atm pressure. What will be the volume when the pressure is changed to 720. torr?

   a) 237 mL   b) 474 mL   c) 528 mL   d) 1.06×10^3 mL

8. At constant temperature the volume of a gas is inversely proportional to its pressure. Which statement correctly expresses this relationship?

   a) $\dfrac{P_1}{V_1} = \dfrac{P_2}{V_2}$ b) $PV = $ constant   c) $\dfrac{P}{V} = $ constant   d) $V = P$

9. A 6.00 L sample of O_2 is at a pressure of 760 torr. What must the pressure be to change the volume to 1.40 L (T is constant)?

   a) 1.77 torr   b) 3.26×10^3 torr   c) 0.233 atm   d) 4.00 atm

10. If the pressure on 125 L of gas is changed from 2.52 atm to 1.50 atm the new volume will be

    a) 210 L   b) 74.4 L   c) 473 mL   d) No correct answer given.

11. What volume will 3.00 L of a gas occupy if the temperature is changed from 30°C to 90°C (P is constant)?

    a) 1.00 L b) 9.00 L c) 2.50 mL d) No correct answer given.

12. Given 10.0 L of N_2 at - 78°C. What volume will the N_2 occupy at 25°C (P is constant)?

    a) 6.54 L b) 18.0 L c) 5.51 L d) 15.3 L

13. The combined gas law includes all of the following laws except

    a) Boyles's Law b) Gay-Lussac's Law c) Charles's Law d) Avogadro's Law

14. The pressure exerted by a mixture of an ideal gas is given by

    a) Boyle's Law b) Charles's Law c) Gay-Lussac's Law d) Dalton's Law

15. Given 0.838 L of NH_3 at 30.0°C. At what temperature will the NH_3 have a volume of 2.00 L (P is constant)?

    a) 127 K b) 450°C c) 71.6°C d) 12.6°C

16. Given 5.00 L of Cl_2 gas at 10°C and 1.00 atmosphere, the volume at 80°C and 2.00 atmosphere will be

    a) 3.12 L b) 20.0 L c) 8.02 L d) 12.5 L

17. A gas occupies 2.88 L at 80.0°C and 725 torr. What volume will it occupy at - 15°C and 780. torr?

    a) 1.96 L b) 3.66 L c) 2.26 L d) 4.24 L

18. A gas has a volume of 125 mL at 630 torr and 27°C. What will be the volume at STP?

    a) 94.3 mL b) 114 mL c) 137 mL d) 166 mL

19. A mixture of gases contains O_2 at 125 torr, N_2 at 125 torr, and H_2 at 450. torr pressure. What is the total pressure of the mixture?

    a) 450. torr b) 525 torr c) 700. torr d) 900. torr

20. A mixture of gases contains 0.200 mole O_2, 0.400 mole N_2, and 0.600 mole H_2 at 2.00 atm pressure. What is the partial pressure of N_2 in the mixture?

    a) 1.00 atm b) 333 atm c) 0.667 atm d) insufficient data to answer

21. What is the partial pressure of O_2 collected over water at 25°C and 630 torr? (Vapor pressure of H_2O at 25°C is 23.8 torr.)

    a) 654 torr b) 681 torr c) 606 torr d) 23.8 torr

22. A mixture of gases containing 1 mole each of He, Ne, and Ar will occupy what volume at STP?

    a) 22.4 L b) 44.8 L c) 67.2 L d) insufficient data to answer

23. One liter of H_2 at STP contains

    a) 1.00 mole H_2 b) $\underline{1.00}$ mole H_2 c) 22.4 mole H_2 d) No correct answer given.
                                 22.4

24. The volume of O_2 collected over water at 25°C and 740 torr was 250 mL. What volume will the dry O_2 occupy at STP? (Vapor pressure of water at 25°C is 23.8 torr.)

    a) 216 mL   b) 242 mL   c) 257 mL   d) 290 mL

25. What volume will 5.50 mol of C_2H_2 (acetylene) occupy at STP?

    a) 4.74 L   b) 4.36 L   c) 246 mL   d) 123 L

26. Which one of the following gases has the lowest density?

    a) Ar   b) CO_2 c) CH_4 d) Kr

27. What volume will 1.00 g of propane (C_3H_8) occupy at STP?

    a) 1.96 L   b) 0.509 L   c) 1.0 L   d) 22.4 L

28. How many grams of tetrafluoroethylene, C_2F_4, are in 2.50 L of C_2F_4 at STP?

    a) 0.112g   b) 896 g   c) 10.0 g   d) 11.2 g

29. If 4.00 L of a gas at STP weighs 5.36 g, its molar mass is

    a) 16.7 g/mole   b) 30.0 g/mole   c) 1.34 g/mole   d) 32.0 g/mole

30. At STP a gas has a density of 1.4 g/L. The numerical value of its molar mass is given by the expression

    a) 1.4 x 22.4   b) $\frac{1.4}{22.4}$ c) $\frac{22.4}{1.4}$ d) 22.4 x 760 x 1.4

31. Of the following gases, H_2, He, CH_4, and Ne, which one at STP has the greatest density?

    a) He   b) H_2 c) CH_4 d) Ne

32. The density of O_2 at STP is the same as the density of

    a) CH_4 at STP   b) CH_4 at standard temperature and 2 atm pressure
    c) SO_2 at STP and 2 atm pressure   d) No correct answer given.

33. What is the density of HBr gas at STP?

    a) 0.361 g/L   b) 0.227 g/L   c) 3.61 g/L   d) 2.27 g/L

34. What is the density of O_2 at 1 atm pressure and 75.0°C?

    a) 1.43 g/L   b) 1.12 g/L   c) 32.0 g/mol   d) 0.894 g/L

35. What volume is occupied by 15.0 g of HCl gas at 715 torr, 90°C?

      a) 13.0 L   b) 9.79 L   c) 475 L   d) 6.51 L

36. What is the volume of 1.00 mole of CH_4 at 273°C and 380 torr?

      a) 44.8 L   b) 89.6 L   c) 11.2 L   d) 22.4 L

37. What pressure is exerted by 0.30 mole of Cl_2 in a 4.00 L container at 20°C?

      a) 94 torr   b) 1.8 atm   c) 8.1 mm Hg   d) 1.7 atm

38. How many moles of Cl_2 gas occupy 10.0 L at - 10°C and 4.50 atm pressure?

      a) 2.08 mol   b) 1.94 mol   c) 1.00 mol   d) 2.22 mol

39. What is the molar mass of a gas, 5.25 g of which occupy 1.25 L at 30.0°C and 0.915 atm pressure?

      a) 28.0 g/mole   b) 114 g/mole   c) 146 g/mole   d) 179 g/mole

40. If 4.0 g of O_2 occupy 2.0 L at a pressure of 1.00 atm the temperature of the O_2 is

      a) 61°C   b) 6.1 K   c) - 78°C   d) 195°C

41. A sample of SO_2 that measures 800. mL at STP will have a mass of

      a) 2.3×10^3 g   b) 51.2 g   c) 12.5 g   d) 2.29 g

The next question set (37 –40) pertains to the following balanced gaseous equation:

      $N_2(g) + 3 H_2(g) \rightarrow 2 NH_3 (g)$ (constant T and P)

42. 3 L of N_2 will react with

      a) 3L H_2 b) 6L H_2 c) 9L H_2 d) 12 L H_2

43. 2.0 L of H_2 will react with

      a) 2.0 L N_2 b) 0.67 L N_2 c) 0.33 L N_2 d) 28 g N_2

44. If 20 L of NH_3 are obtained, the total volume of reactants reacted was

      a) 10 L   b) 20 L   c) 30 L   d) 40 L

45. The volume of NH_3 obtainable from 6 L of N_2 and 6 L H_2 is

      a) 4 L   b) 6 L   c) 12 L   d) 2 L

46. $C_8H_{16}(g) + 12\ O_2(g) \rightarrow 8\ CO_2(g) + 8\ H_2O(g)$

When 10.0 L of C_8H_{16} are burned, how many liters of water vapor are formed? (P and T constant)

    a) 10.0 L   b) 22.4 L   c) 80.0 L   d) 179 L

47. $C_8H_{16}(g) + 12\ O_2(g) \rightarrow 8\ CO_2(g) + 8\ H_2O(g)$

How many liters of CO_2 at STP are produced when 112.2 g of C_8H_{16} are burned?

    a) 10 L   b) 22.4 L   c) 80.0 L   d) 179 L

48. $Cu(s) + 4\ HN03\ (aq) \rightarrow Cu(NO_3)_2(aq) + 2\ NO_2(g) + 2\ H_2O(g)$

How many liters of NO_2 gas at STP are produced when 16.0 g of Cu are reacted with nitric acid?

    a) 5.6 L   b) 11.3 L   c) 22.4 L   d) 33.6 L

49. $C_3H_8 + 8\ Cl_2 \rightarrow C_3Cl_8 + 8\ HCl$

How many grams of chlorine at STP will react with 16.0 L (at STP) of propane, C_3H_8?

    a) 50.7 g   b) 203 g   c) 405 g   d) 795 g

50. $2\ C_4H_{10}(g) + 13\ O_2(g) \rightarrow 8\ CO_2(g) + 10\ H_2(g)$

What is the percent yield if 65 L of CO_2 were recovered by burning 20. L of C_4H_{10}? (P and T constant)

    a) 25%   b) 81%   c) 91%   d) 100%

51.   At any given temperature and pressure, the rate at which a gas diffuses is inversely proportional to the

    a) molar mass   b) volume
    c) square root of its molar mass   d) square root of its volume

52.   When a gas is heated under constant pressure, the gas molecules

    a) become larger and take up more space
    b) move faster and occupy more space
    c) impact one another with greater force but occupy the same space
    d) repel each other and take up more space

53.   Each of the following would double the volume of a gas *except*

    a) double the pressure   b) double the absolute temperature
    c) halve the pressure   d) double the number of molecules in the container

54.   Which of the following is not equal to 1.00 atm pressure?

    a) 760. cm Hg   b) 29.9 in. Hg   c) 760. mm Hg   d) 760. torr

55. The volume of a dry gas is 4.00 L at 15°C and 745 torr. What volume will the gas occupy at 40°C and 700. torr?

    a) 4.63 L   b) 3.46 L   c) 3.92 L   d) 4.08 L

56. In an effusion experiment, an unknown gas X effuses with a rate of 0.057 m/s. Under the same condition nitrogen, N_2, effuses at a rate of 0.080 m/s. What is the molar mass of the X?

    a) 32 g/mole   b) 48 g/mole   c) 64 g/mole   d) 44 g/mole

57. Which one of the following gases has the slowest effusion rate?

    a) Ar   b) CO_2 c) CH_4 d) Kr

58. The rate of diffusion of H_2 would be how many times faster than the rate of diffusion of CO_2?

    a) 2.2   b) 5.6   c) 6.6   d) 4.2

59. A gas X effuses through a pinhole 1.60 time faster than CO_2 effuses through the same pinhole. Calculate the molar mass of the gas X.

    a) 16.0 g/mole   b) 17.2 g/mole   c) 2.00 g/mole   d) 14.0 g/mole

60. A sample of Cl_2 occupies 8.50 L at 80°C and 740. mm Hg. What volume will the Cl_2 occupy at STP?

    a) 10.7 L   b) 6.75 L   c) 11.3 L   d) 6.40 L

61. What volume will 8.00 g of O_2 occupy at 45°C and 2.00 atm?

    a) 0.462 L   b) 104 L   c) 9.62 L   d) 3.26 L

62. The density of NH_3 gas at STP is

    a) 0.759 g/mL   b) 0.759 g/L   c) 1.32 g/mL   d) 1.32 g/L

63. Box A contains O_2 (molar mass = 32.0) at a pressure of 200 torr. Box B, which is identical to Box A in volume, contains twice as many molecules of CH_4 (molar mass = 16.0) as the molecules of O_2 in Box A. The temperatures of the gases are identical. The pressure in Box B is

    a) 100 torr   b) 200 torr   c) 400 torr   d) 800 torr

64. How many liters of butane vapor are required to produce 2.0 L of CO_2 at STP?
$$2 \, C_4H_{10}(g) + 13 \, O_2(g) \rightarrow 8 \, CO_2(g) + 10 \, H_2O(g)$$
    Butane

    a) 2.0 L   b) 4.0 L   c) 0.80 L   d) 0.50 L

65. What volume of CO_2 (at STP) can be produced when 15.0 g of C_2H_6 and 50.0 g of O_2 are reacted?
$$2 \, C_2H_6(g) + 7 \, O_2(g) \rightarrow 4 \, CO_2(g) + 6 \, H_2O(g)$$

    a) 20.0 L   b) 22.4 L   c) 35.0 L   d) 5.6 L

66. How many molecules are present in 0.025 mol of H_2 gas?

    a) 1.5×10^{22} molecules   b) 3.37×10^{23} molecules
    c) 2.40×10^{25} molecules   d) 1.50×10^{22} molecules

67. 5.60 L of a gas at STP has a mass of 13.0 g. What is the molar mass of the gas?

    a) 33.2 g/mol   b) 26.0 g/mol   c) 66.4 g/mol   d) 52.0 g/mol

68. An ideal gas fills a volume of 31.6 liters at room temperature (25°C) and at a pressure of 73.5 cm Hg. Which expression correctly shows the calculation that would determine the volume of the same gas at a pressure of 765 mm Hg and a temperature of 50.0°C?

    a) $(31.6L)\left(\frac{765}{735}\right)\left(\frac{323}{298}\right)$ b) $(31.6L)\left(\frac{765}{735}\right)\left(\frac{298}{323}\right)$

    c) $(31.6L)\left(\frac{735}{765}\right)\left(\frac{298}{323}\right)$ d) $(31.6L)\left(\frac{735}{765}\right)\left(\frac{323}{298}\right)$

69. Aluminum metal will react with dilute sulfuric acid, H_2SO_4 according to the following equation:

$$2\ Al(s) + 3\ H_2SO_4(aq) \rightarrow 3\ H_2(g) + Al_2(SO_4)_3$$

When 18 g of Al react, how many liters of hydrogen gas at STP will result?

    a) 11.2 L   b) 22.4 L   c) 67.2 L   d) 134 L

70. According to the Kinetic-Molecular Theory

    a) the absolute temperature of a gas depends on its molar mass
    b) gaseous molecules are moving continuously and randomly, and their collisions are perfectly elastic
    c) the pressure exerted on a gas affects the speed of its molecules
    d) gaseous molecules travel in curved paths

71. The density of argon gas at 100°C and 5.0 atmospheres of pressure is

    a) 9.0 g/liter   b) 6.6 g/liter   c) 0.050 g/liter   d) 0.0086 g/liter

72. Gases have a tendency to exhibit real (nonideal) behavior under the conditions of

    a) low temperature and high pressure   b) high temperature and high pressure
    c) low temperature and low pressure   d) any temperature above absolute temperature

73. If 6.6 g of a gaseous compound occupy a volume of 1.20 liters at 27°C and 0.967 atm, the molar mass of that gas must be

    a) 165 g/mol   b) 140 g/mol   c) 123 g/mol   d) 109 g/mol

74. Which of the following is true regarding ozone, O_3?

    a) O_2 and O_3 are allotrope
    b) O_3 absorbs harmful ultraviolet radiation from the sun
    c) O_3 can cause extensive damage to plants
    d) All are true.

75. Which of the following gases contribute to urban air pollution?

    a) CO_2 b) SO_2 c) NO_2 d) All contribute to pollution.

CHAPTER 13. Water and the Properties of Liquids

True — False (choose one)

1. Water covers about 70% of the earth's surface.

2. Water is the most abundant compound in the human body.

3. Water is a colorless, odorless, tasteless liquid.

4. The heat of fusion is the amount of heat energy required to change a solid to a liquid at its melting point.

5. The heat of fusion of water is 80 J/g or 355 cal/g.

6. Heat of vaporization is the amount of heat energy required to change one gram of liquid to a vapor at its boiling point.

7. Water always boils at 100°C.

8. Steam at 100°C contains the same amount of heat energy per gram as liquid water at 100°C.

9. Water has its maximum density at 4°C.

10. Water has its maximum density just before it freezes.

11. Water has the highest melting point of the Periodic Group VIA hydrogen compounds (H_2O, H_2S, H_2Se, H_2Te).

12. Water molecules are nonpolar.

13. A water molecule is a linear molecule.

14. The hydrogen oxygen bond in water is a nonpolar covalent bond.

15. Fluorine, the most electronegative element, forms the strongest hydrogen bonds.

16. Hydrogen bonds hold the hydrogen atoms together in a hydrogen molecule.

17. Hydrogen bonding accounts for the relatively high boiling point and freezing point of pure water.

18. Hydrogen bonds in water are intramolecular bonds.

19. A hydrogen bond is a chemical bond that is formed between polar molecules that contain hydrogen covalently bonded to a small electronegative atom such as fluorine, oxygen, or nitrogen.

20. Hydrogen bonds are stronger than ionic and covalent bonds.

21. Evaporation is the escape of molecules from the liquid state to the gas or vapor state.

22. Water, which has a molar mass of 18.02 g/mol and boils at 100°C, will evaporate faster than ethyl alcohol which has a molar mass of 46.07 g/mol and boils at 78°C.

23. Sublimation is a form of evaporation in which a substance goes directly from the solid state to the gaseous state.

24. Surface tension of a liquid is the tendency of the molecules at the surface of a liquid to be pulled inward.

25. Surface tension decreases with increasing temperature.

26. Boiling point and melting point are generally used to characterize and identify substances.

27. A liquid has an infinite number of boiling points given by the vapor pressure curve.

28. As the temperature of a liquid is increased, its vapor pressure decreases.

29. At pressures below atmospheric (760 torr), water boils below 100°C.

30. The vapor pressure of a liquid is dependent on the temperature and the amount of liquid and vapor present.

31. When equal volumes of two liquids are placed in two separate beakers, the lower molar mass liquid will evaporate faster.

32. At the same temperature all liquids have the same vapor pressure.

33. A volatile liquid will have a low vapor pressure.

34. We have two liquids at room temperature. The liquid with the higher vapor pressure will have the lower boiling point.

35. The vapor pressure of ethyl alcohol at its normal boiling point is 760 torr.

36. Sulfur dioxide, SO_2, boils at - 10°C. The vapor pressure of boiling SO_2 is 760 torr.

37. The boiling point of H_2O is the temperature at which the vapor pressure of the water equals the prevailing atmospheric pressure.

38. The melting point of a substance is the temperature at which the solid substance is in equilibrium with its liquid.

39. The reaction of hydrogen with oxygen is exothermic.

40. The fact that water reacts with many substances indicates it is an unstable substance.

41. Heating water at its boiling point decomposes water into the gases hydrogen and oxygen.

42. When water is decomposed, the ratio of the gases formed is one volume of hydrogen to two volumes of oxygen.

43. Water is produced as a product of acid-base neutralization reactions.

44.   Sodium, potassium, and calcium react with cold water to produce hydrogen and a base.

45.   Iron reacts with steam (H_2O) according to the following equation:

$$2 \text{ Fe(s)} + 2 \text{ H}_2\text{O(g)} \rightarrow 2 \text{ H}_2\text{Fe(s)} + \text{O}_2\text{(g)}$$

46.   Copper, silver, and gold do not react with water to produce hydrogen gas.

47.   A metal oxide that reacts with water to form a base is known as a metal anhydride.

48.   A nonmetal oxide that reacts with water to form an acid is known as an acid anhydride.

49.   The anhydride of $Ba(OH)_2$ is BaO_2.

50.   The anhydride of H_2SO_4 and HNO_3 are SO_3 and NO_2, respectively.

51.   A hydrate is an ionic compound in which the formula unit includes a fixed number of water molecules together with cations and anions.

52.   A gas hydrate is a gas molecule trapped inside the crystal lattice of water molecules.

53.   The anhydride of H_2SO_3 is SO_3.

54.   When SO_2 gas is dissolved in water, the resulting solution will be acidic.

55.   When K_2O is dissolved in water, the resulting solution will be acidic.

56.   MgO is an acid anhydride.

57.   MgO is a basic anhydride.

58.   Solids that contain water molecules as part of their crystalline structure are known as hydrides.

59.   Water of crystallization may be removed from a compound by moderate heating.

60.   The name for $FeCl_3 - 6H_2O$ is iron(III) chloride hexahydrate.

61.   The name for $NaC_2H_3O_2 - 3H_2O$ is sodium carbonate trihydrate.

62. $CoCl_2 - 6H_2O$ has a higher percent of water than does $CaCl_2 - 6H_2O$.

63.   One mole of the hydrate $Ba(OH)_2 - 8H_2O$ contains eight moles of water.

64. $CaCl_2 - 2H_2O$ contains 14.7% water of hydration.

65.   Most natural fresh waters are safe to drink.

66.   An optimum amount of fluoride in drinking water is beneficial in that it makes the teeth of children more resistant to decay.

67. Water that contains dissolved calcium and magnesium ions is called "hard" water.

68. A major drawback to the use of "hard" water is that it makes cooked foods hard to digest.

69. Sodium, potassium, and calcium each react with cold water to form hydrogen gas and a metal hydroxide.

70. Water in a hydrate is known as water of hydration or water of crystallization.

71. The pressure exerted by a vapor in equilibrium with its liquid is known as the vapor pressure of the liquid.

72. The heat of vaporization of water is more than six times as large as its heat of fusion.

73. Of two liquids, the one having the higher vapor pressure at a given temperature will have the higher normal boiling point.

74. In treating city water supplies, fluorides are often added to prevent tooth decay.

75. The molar heat of fusion of water is 335 J/g x 18.0 g/mol.

76. In treating city water supplies, chlorine is injected into the water to kill harmful bacteria before it is distributed to the public.

77. Metal oxides that react with water to form bases are known as basic anhydrides.

78. Nonmetal oxides that react with water to form acids are known as acid anhydrides.

79. When P_2O_5 reacts with water, it would be expected to make the solution basic.

80. Hydrogen bonds are stronger than covalent bonds.

81. The bond angle between the atoms in a water molecule is about 90°.

82. Capillary action results from the adhesive forces within a liquid and the cohesive forces between the liquid and the walls of the container.

83. When mercury is placed in a glass tube, the meniscus will be convex.

84. A substance with a high value for boiling point is often also a substance with a high heat of vaporization.

85. The boiling temperature and the vapor pressure of any liquid are higher at higher altitudes than lower altitude.

86. If 1000 J of heat energy are added to 10 g of ice at 0°C, the ice will completely melt to liquid water.

87. CO_2 is a more acidic anhydride than CaO.

88. More heat is required to vaporize 1 g of liquid water than to melt 1 mole of ice.

Use the following heating curve to answer the next set of questions.

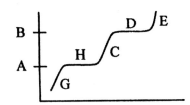

89.   Point D represents the melting process of a substance.

90.   Point A represents the boiling point of a substance.

91.   Point E represents the gaseous phase of a substance.

Multiple Choice (**choose the best answer**)

1.   Water is a bent molecule, $H-O-H$. The bond angle between hydrogen atoms is about

        a) 90°   b) 105°   c) 135°   d) 180°

2.   The number of joules required to change 1 g of ice at 0°C to 1 g of steam (gas) at 100°C is

        a) 2.26 kJ   b) 335 J   c) 418 J   d) sum of a, b, and c.

3.   The heat of fusion of water is

        a) 4.184 J/g   b) 2.26 kJ/g   c) 335 J/g   d) 2.26 kJ/mol

4.   The heat of vaporization of water is

        a) 4.184 J/g   b) 2.26 kJ/g   c) 335 J/g   d) 2.26 kJ/mol

5.   The specific heat of water is

        a) 4.184 J/g°C   b) 2.26 kJ/g°C   c) 335 J/g°C   d) 2.26 kJ/mol°C

6.   How many kilojoules are required to change 75.0 g of water at 25.0°C to steam at 100°C?

        a) 170. kJ   b) 193 kJ   c) 23.5 kJ   d) No correct answer given.

7.   How many calories are required to change 25.0 g of water at 25°C to steam at 100°C?

        a) 1.9×10^3 cal   b) 1.4×10^4 cal
        c) 1.5×10^4 cal   d) No correct answer given.

8.   How many joules of energy must be removed to change 45 g of water at 25°C to ice at 0°C?

   a) 2.0×10^4 J   b) $1.5 \times 10_4$ J   c) 4.7×10^3 J   d) 8.4×10^3 J

9.   The effect of hydrogen bonding is greatest when hydrogen is

   a) covalently bonded to an element of high electronegativity
   b) covalently bonded to an element of low electronegativity
   c) in the form of its positive ion, H+
   d) in the form of its negative ion, H-

10.   Which of these elements has the greatest hydrogen bonding?

   a) Oxygen   b) Fluorine   c) Nitrogen   d) Sodium

11.   Which of the following properties of water is not affected by hydrogen bonding?

   a) boiling point   b) freezing point   c) vapor pressure   d) molar mass

12.   In which compound is hydrogen bonding not important?

   a) HF   b) NH_3 c) CH_3OH d) HBr

13.   Which compound does not exhibit hydrogen bonding?

   a) NH_3 b) CH_4 c) H_2O d) HF

14.   Which of the following structures exhibits the least hydrogen bonding?

   a) F-H   b) -O-H   c) N-H   d) Na-H

15.   Water boils at 100°C; ethyl alcohol boils at 78°C. Which compound will have the higher vapor pressure at their respective boiling point?

   a) water   b) ethyl alcohol
   c) neither - both have the same vapor pressure   d) insufficient data to answer

16.   Liquid A boils at a higher temperature than liquid B. This fact indicates that

   a) A has a higher vapor pressure than B at any particular temperature
   b) A has a lower vapor pressure than B at any particular temperature
   c) the molar mass of A is greater than B
   d) No correct answer given.

17.   Which one of the following solution properties is independent of temperature?

   a) vapor pressure   b) surface tension
   c) heat of fusion   d) All the above depend on temperature.

18. Which one of the following properties has a relatively high value in a liquid with weak attractive force?

    a) vapor pressure   b) surface tension   c) heat of vaporization   d) boiling point

19. Which one of the following properties has a relatively low value in a liquid with strong attractive force?

    a) vapor pressure   b) surface tension   c) heat of vaporization   d) boiling point

20. Which one of the following properties has a relatively high value in a liquid with strong attractive force?

    a) boiling point   b) surface tension   c) heat of vaporization   d) All of the above.

21. At which temperature is the vapor pressure of water the greatest?

    a) at its freezing point
    b) at its normal boiling point
    c) at 350 K
    d) at its boiling point when the atmospheric pressure is 380 torr

The next question set (22 – 27) refers to this statement: The vapor pressures of 4 liquids, A, B, C, and D, at 50°C are, respectively, 200 torr, 178 torr, 500 torr and 60 torr.

22. As the pressure over the liquids is reduced which liquid will boil first?

    a) A   b) B   c) C   d) D

23. Which liquid will have the highest normal boiling point?

    a) A   b) B   c) C   d) D

24. Which liquid is the most volatile?

    a) A   b) B   c) C   d) D

25. Attractive forces are greatest between molecules of which liquid?

    a) A   b) B   c) C   d) D

26. Which liquid will be most effective in cooling by evaporation?

    a) A   b) B   c) C   d) D

27. Equal volumes of the 4 liquids are placed in separate uncovered 250 mL beakers and let stand until one liquid has completely evaporated. The three liquids remaining and their relative amounts (from least to most) are

    a) C, A, B   b) B, A, C   c) D, B, A   d) A, B, D

28. Which process is not endothermic?

    a) evaporation   b) sublimation   c) boiling   d) crystallization

29. Mercury is a non-polar liquid with strong attractive (cohesive) force. The meniscus of mercury placed in a glass cylinder is expected to be

    a) concave   b) convex   c) flat   d) oscillates between concave and convex

30. Which one of the followings statements is *not* correct?

    a) The equilibrium vapor pressure above a liquid is independent of the volume of liquid present.
    b) The stronger the attractive force, the higher the boiling point.
    c) A volatile liquid has a high boiling point.
    d) The vapor pressure above a liquid increases with increasing temperature.

31. The reason a water strider can walk on water is due to

    a) capillary action of water   b) viscosity of water
    c) surface tension of water   d) vapor pressure of water

32. Which one of the following is incorrect regarding a hydrate?

    a) Water molecules are part of the crystalline lattice.
    b) Water molecules can be removed by moderate heating of the crystal.
    c) Water molecules are bonded by covalent forces.
    d) Water molecules are bonded by electrostatic forces.

33. The best term that describes the raise of water up a glass cylinder is

    a) capillary action   b) surface tension   c) cohesive force   d) vapor pressure

This heating curve (below) for a particular substance pertains to the following question set:

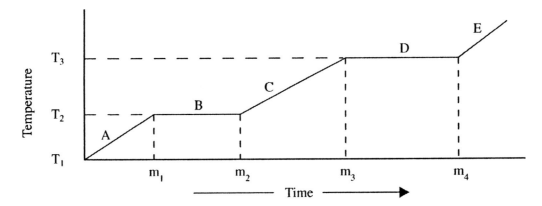

34. What is the phase of the substance in the portion of the curve labeled C?

    a) solid   b) liquid   c) gas   d) insufficient data to determine

35. T_2 is the temperature at which the substance

    a) boils   b) melts   c) vaporizes   d) sublimes

36. The substance begins to boil at the time labeled

    a) m_1 b) m_2 c) m_3 d) m_4

37. In which portion of the curve do the molecules have the greatest kinetic energy?

    a) B   b) A   c) E   d) D

38. The portion of the curve labeled D represents the time during which the substance is

    a) being warmed as a solid   b) being warmed as a liquid
    c) changing from solid to liquid   d) changing from a liquid to a vapor at its boiling point

39. Which compound is not an acid anhydride?

    a) CO_2 b) N_2O_5 c) Na_2O d) Cl_2O_7

40. Which compound is a basic anhydride?

    a) NaCl   b) CH_4 c) CaO   d) $Al(OH)_3$

41. Which of these metals reacts most vigorously with H_2O?

    a) Na   b) K   c) Ca   d) Zn

42. The anhydride of H_3PO_4 is

    a) PO_4 b) PO_3 c) P_2O_5 d) P_2O_3

43. The anhydride of HNO_3 is

    a) N_2O_5 b) N_2O_3 c) NO_2 d) NO

44. The anhydride of NH_4OH is

    a) N_2O b) NO   c) NH_3 d) N_2H_4

45. The anhydride of $Fe(OH)_3$ is

    a) FeO   b) Fe_2O_3 c) Fe_3O_4 d) Fe_2O

46. The anhydride of LiOH is

    a) Li_2O b) LiO   c) Li_2O_3 d) LiO_2

47. Which of the following is a neutralization reaction?

    a) $CaO + H_2O \rightarrow Ca(OH)_2$ b) $CH_4 + 2\,O_2 \rightarrow CO_2 + 2\,H_2O$
    c) $2\,KOH + H_2SO_4 \rightarrow K_2SO_4 + 2\,H_2O$ d) $H_2SO_4 + BaCl_2 \rightarrow BaSO_4 + 2HCl$

48. Which is an incorrect prefix for indicating the number of water molecules in a hydrate?

    a) bi   b) tetra   c) tri   d) hexa

49. Which hydrate has the highest percent of water?

    a) $NiCl_2 - 6H_2O$ b) $CoCl_2 - 6H_2O$ c) $MgCl_2 - 6H_2O$ d) $CaCl_2 - 6H_2O$

50. A technique used to soften hard water is

    a) distillation   b) chemical precipitation   c) ion exchange   d) a, b, and c

51. The boiling point of pure water

    a) does not depend upon location or elevation
    b) depends on atmospheric pressure
    c) is always higher at higher elevations than at lower ones
    d) is higher than a solution of salt water

52. The change in the volume of a liquid as it freezes to a solid is always much less than the decrease in volume of a gas as it becomes a liquid. This is because

    a) liquid molecules repel each other more
    b) gas molecules are much more widely separated than liquid molecules
    c) gas molecules are attracted to each other
    d) liquid molecules are much larger than gas molecules

53. Suppose you have a closed container at a constant temperature with only a liquid inside. Which is true?

    a) The vapor pressure will increase as gas molecules come into existence.
    b) The vapor pressure will not change.
    c) The vapor pressure will decrease as the liquid stays intact.
    d) The vapor pressure decreases as gas molecules come into existence.

54. The specific heat of water is

    a) 4.184 J/g °C   b) 335 J/g °C   c) 2.26 kJ/g °C   d) 18 J/g °C

55. The density of water at 4°C is

    a) 1.0 g/mL   b) 80 g/mL   c) 18.0 g/mL   d) 14.7 lb/in^3

56. In which of the following molecules will hydrogen bonding be important?

$$a)\ H\!-\!F \quad b)\ \begin{matrix} S\!-\!H \\ | \\ H \end{matrix} \quad c)\ H\!-\!Br \quad d)\ \begin{matrix} H & & H \\ | & & | \\ H\!-\!C\!-\!O\!-\!C\!-\!H \\ | & & | \\ H & & H \end{matrix}$$

57. Which of the following is an incorrect equation?

a) $C + H_2O(g) \xrightarrow{1000°} CO(g) + H_2(g)$ b) $CaO + H_2O \rightarrow Ca(OH)_2$

c) $2 NO_2 + H_2O \rightarrow 2 HNO_3$ d) $Cl_2 + H_2O \rightarrow HCl + HOCl$

58. The formula for iron (H) sulfate heptahydrate is

a) $Fe_2SO_4 - 7 H_2O$ b) $Fe(SO_4)_2 - 6 H_2O$ c) $FeSO_4 - 7 H_2O$ d) $Fe_2(SO_4)_3 - 7 H_2O$

59. The process by which a solid changes directly to a vapor is called

a) vaporization   b) evaporation   c) sublimation   d) condensation

60. The anhydride of permanganic acid, $HMnO_4$, is

a) MnO_3 b) Mn_2O_3 c) MnO_4 d) Mn_2O_7

61. Consider two beakers, one containing 50 mL of liquid A and the other 50 mL of liquid B. The boiling point of A is 90°C and that of B is 72°C. Which of these statements is correct?

a) A will evaporate faster than B   b) B will evaporate faster than A
c) both A and B evaporate at the same rate   d) insufficient data to answer the question

62. Water can exist at conditions of 100°C and 1.00 atmosphere pressure as

a) only a gas   b) only a liquid   c) a liquid and a gas   d) a liquid and a solid

63. The unusually high heat of vaporization of water is mainly a result of

a) ionic bonds   b) covalent bonds   c) dipole forces   d) hydrogen bonds

64. The normal boiling point of a liquid is defined to be

a) the pressure at which a liquid vaporizes
b) the temperature at which the vapor pressure of a liquid equals 1 atm
c) the temperature at which the vapor pressure of a liquid equals the atmospheric (barometric) pressure
d) the temperature at which a liquid vaporizes

65. Which equation correctly shows the reaction between elemental potassium and water?

a) $2K + H_2O \rightarrow H_2 + 2KOH$ b) $2K_2 + 2H_2O \rightarrow 4HK + O_2$
c) $K + H_2O \rightarrow H_2KO$ d) $2K + H_2O \rightarrow HK + KOH$

66. The vapor pressure of a liquid in a container does not depend upon

a) the type of liquid it is
b) the temperature of the liquid
c) the intermolecular forces between the liquid molecules
d) the atmospheric pressure in the room

Matching

From the list of terms given choose the one that correctly identifies each phrase.

List of terms: (a) normal boiling point; (b) sublimation; (c) deliquescence; (d) basic anhydride; (e) acid anhydride; (f) water; (g) boiling point temperature; (h) melting point temperature; (i) vapor pressure; (j) evaporation; (k) hydrate

_____ 1.   Nonmetal oxides that react with water to form acids

_____ 2.   Solids that contain water molecules as part of their crystalline structure

_____ 3.   Metal oxides that react with water to form bases

_____ 4.   Substance which covers 70% of the earth's surface

_____ 5.   The process by which molecules go from a liquid state to a gas or vapor state

_____ 6.   The process by which a solid passes directly to the gaseous state without liquefying

_____ 7.   The pressure exerted by a vapor in equilibrium with its liquid

_____ 8.   The temperature at which the vapor pressure of a liquid is equal to the external pressure above the liquid

_____ 9.   The temperature at which the solid phase of a substance is in equilibrium with its liquid phase

CHAPTER 14.   Solutions

True — False (choose one)

1.   A solution is a homogeneous mixture.

2.   The solute is the most abundant component of a solution.

3.   A solution can contain more than one solute and more than one solvent.

4.   A solution is always composed of a solid solute and a liquid solvent.

5.   The dissolving medium in solutions is always a liquid.

6.   Both the solute and solvent in a solution may be a solid, a liquid, or a gas.

7.   A solution has a fixed composition of solute and solvent.

8.   Solutions are clear or transparent, even though they can have a characteristic color.

9.   A true solution cannot be cloudy.

10.   An insoluble solute dissolved in a solvent cannot form a solution.

11.   If two liquids are miscible a mixture of the two will be homogeneous.

12.   Nonpolar substances tend to be soluble in water.

13.   Polar substances tend to be soluble in water.

14.   Pulverizing a solid to small pieces will increase the rate of dissolving of the solid in a liquid.

15.   Two liquids which are not capable of mixing and forming a solution are said to be miscible.

16.   All the following solids are soluble in water: NH_4NO_3, KCl, $CuSO_4$, Na_2CO_3, $Ba(NO_3)_2$, NaOH and $KC_2H_3O_2$.

17.   All the following solids are soluble in water: NaI, KBr, NH_4Br, $BaSO_4$, $AgNO_3$, and CuS.

18.   The solubility of a gas in a liquid is increased as the pressure of that gas above the liquid is increased.

19.   Solubility is a colligative property.

20.   An increase in particle size will increase the rate of dissolving of a solid in a liquid.

21.   An increase in temperature increases the solubility of most solids.

22.   An increase in temperature almost always increases the rate at which a solid will dissolve in a liquid.

23.   The more concentrated a solution is, the faster it will dissolve additional solute.

24. If only a small amount of solute dissolves in a solvent, it would be correct to call that solution dilute *and* saturated.

25. When a solute and solvent are first mixed the rate of dissolving is at a maximum.

26. A dilute solution contains more solute than a concentrated solution.

27. A saturated solution contains dissolved and undissolved solute in equilibrium.

28. When a saturated solution of KNO_3 is cooled, solid KNO_3 will precipitate out of solution.

29. Dilute and concentrated are quantitative expressions of solution concentration.

30. Dilute aqueous solutions have molarities and molalities that are nearly the same.

31. Addition of pure solvent to a solution lowers the molarity of the solution and consequently, the moles of solute present.

32. A 50% NaCl solution has the same density as a 50% $CaCl_2$ solution.

33. The mass percent of a solution containing 12 g of $NaNO_3$ in 50 g of water is 24%.

34. 150 mL of 70% alcohol by volume contains 105 mL of the alcohol.

35. If 40 mL of benzene is added to 30 mL of toluene the volume percent of benzene in the mixture is 57%.

36. If 2.0 g of sucrose is added to 6.0 g of water, the mass percent of sucrose in the solution is 25%.

37. A 1 molar solution contains 1 mole of solute per liter of solution.

38. A 0.5 molar solution contains 0.5 mole of solute per 500 mL of solution.

39. 500 mL of 2.50 M NaOH solution contain 5.00 mol of NaOH.

40. 500 mL of 2.50 M NaOH solution contain 1.25 mol of NaOH.

41. There are 2.0 moles of $NaC_2H_3O_2$ in 500 mL of 4.0 M solution.

42. To prepare a 1.0 M NaOH solution, dissolve 0.10 mol of NaOH in water and dilute the solution to 100 mL.

43. A 1 molar solution contains 1 mole of solute dissolved in 1000 g of solvent.

44. Freezing point depression, boiling point elevation, vapor pressure lowering, and molarity are colligative properties of solutions.

45. The colligative properties of solutions are directly proportional to the molality of the solute particles in solution.

46. When a solute is dissolved in a solvent, the freezing point of the solution will be higher than that of the pure solvent.

47. A solution made by dissolving 0.200 mol of ethylene glycol, $C_2H_6O_2$, in 100 g of water is a 2.00 molal solution.

48. A solution made from 100 g of ethanol, C_2H_5OH, and 100 g of water will freeze at a lower temperature than a solution of 100 g of methanol, CH_3OH, and 100 g of water.

49. Osmotic pressure, freezing point depression, vapor pressure lowering, and boiling point elevation all depend on the number of solute particles in a given quantity of solution.

50. A solution of 0.200 mole of sucrose and 200 g of water will freeze at -1.86°C.

51. The molality of a solution is independent of temperature.

52. Carbonated beverages are solutions with a liquid solvent and a gaseous solute.

53. Whipped cream is a solution with a liquid solvent and a liquid solute.

54. A drinking glass of water, upon standing, reveals bubbles in the water. This gives evidence of a gaseous solute in the water.

55. A 50% solution will always contain 50 g of solute.

56. Solubility describes the amount of one substance that will dissolve into another substance.

57. Osmosis is the net flow of water through a semipermeable membrane from the region of low molarity to the region of high molarity.

58. At a given temperature, the molarity of a supersaturated solution is lower than the molarity of a saturated solution.

59. Molality, percent by mass, and mole fraction are all temperature-independent measures of concentration.

60. When solid NaCl dissolves in water, the sodium ions attract the positive end of the water dipole.

61. For most solids or gases dissolved in liquid, an increase in temperature results in an increase in solubility.

62. A supersaturated solution contains dissolved solute in equilibrium with undissolved solute.

63. Liquids that do not mix are said to be miscible.

64. As a general rule, sodium and potassium salts of the common ions are soluble in water.

65. The vapor pressure of the solvent in a solution is lower than the vapor pressure of the pure solvent.

66. A solution containing a nonvolatile solute has a lower boiling point than the pure solvent as a result of having a lower vapor pressure than the pure solvent.

67. A major use of a solvent is as a medium for chemical reactions.

68. Molar solutions are temperature dependent, because the volume of the solution can vary with the temperature but the number of moles of solute remains the same.

69. A 15% by mass solution contains 15 g of solute per 100 mL of solution.

70. A 1 M solution of NaCl and a 1 M solution of $NaNO_3$ would have the same densities.

71. A 1 M solution of NaCl and a 1 M solution of $NaNO_3$ would have the same number of solute particles in solution.

72. All solutes dissociate into ions once they are dissolved in water.

73. It is not possible to make a solution using a liquid as a solvent and a gas as a solute.

74. Ionic compounds that are salts of sodium are insoluble in water.

75. Lead (II) nitrate is insoluble in water.

76. Commercial sulfuric acid has a density of 1.84 g/ml and is 96.0% H_2SO_4 by mass. If 25.0 mL of this concentrated acid is diluted to 500. mL with water, the molarity of the final solution is 0.900 M.

77. Suppose that both Ca^{2+} and Pb^{2+} ions are present in a solution. If 3 M HCl is added to the solution, both $CaCl_2$ and $PbCl_2$ will precipitate.

Multiple Choice (choose the best answer)

1. Which pair of liquids is miscible?

    a) H_2O – acetic acid   b) H_2O – CCl_4 c) oil – H_2O d) gasoline – H_2O

2. Which class of compound is generally soluble in H_2O?

    a) carbonates   b) phosphates   c) nitrates   d) hydroxides

3. Substances that are capable of mixing and forming a solution are

    a) saturated   b) miscible   c) supersaturated   d) dilute

4. According to the general solubility rules, which one of the following compounds would be insoluble?

    a) NaCl   b) FeS   c) $CuBr_2$ d) $(NH_4)_2S$

5. According to the general solubility rules, which one of the following compounds would be soluble?

    a) $Cr(NO_3)_3$ b) $CaCO_3$ c) AgI   d) $Ba_3(PO_4)_2$

6. If NaCl is soluble in water to the extent of 36.0 g NaCl/100 g H_2O at 20°C, then a solution at 20°C containing 45 g NaCl/150 g H_2O would be

    a) dilute   b) saturated   c) supersaturated   d) unsaturated

7. The solubility of sodium nitrate ($NaNO_3$) in water varies with the

   a) amount of $NaNO_3$ b) the amount of water
   c) pressure of the air   d) temperature of the water

8. Which procedure is most likely to increase the solubility of solids in liquids?

   a) stirring the solution   b) pulverizing the solid
   c) heating the solution   d) increasing the pressure

9. A saturated solution of sodium nitrate contains 80 g $NaNO_3$ per 100 g H_2O. How many grams of $NaNO_3$ must be added to a solution containing 25 g $NaNO_3$ in 75 g H_2O to make the solution saturated?

   a) 60 g   b) 45 g   c) 35 g   d) 15 g

10. A change in pressure would have the greatest effect in the solubility of a

    a) solid in a solid   b) gas in a liquid
    c) liquid in a liquid   d) liquid in a solid

11. A student dissolved 100 g of a yellow powder in 500 mL of solvent. The concentration of the solution is

    a) 100 g solute/500 mL solvent   b) 100 g solvent/500 mL solute
    c) 100 g solute/600 g solution   d) 100 g solute/600 ml solution

12. The addition of one crystal of $NaClO_3$ to a solution of this salt causes more crystals to precipitate from solution. If there is no temperature change, the resulting solution is

    a) dilute   b) unsaturated   c) supersaturated   d) saturated

13. If 15.0 g of KNO_3 is added to 75 g of water, what is the mass percent of KNO_3 in the solution?

    a) 16.7%   b) 20.0%   c) 19.0%   d) 0.165%

14. The mass percent of a solution prepared by dissolving 0.500 mol of KCl in 850 g of water is

    a) 4.20%   b) 8.07%   c) 5.55%   d) 3.60%

15. 100 mL of solution contain 50.0 g of NH_4NO_3 and has a density of 1.20 g/mL. The mass percent of the solution is

    a) 33.3%   b) 29.4%   c) 41.7%   d) 50.0%

16. A solution is made by dissolving 1.0 mol of methanol, CH_3OH, in 1.0 mol of water. What percent of the molecules in the solution are methanol?

    a) 50%   b) 63%   c) 38%   d) 76%

17. What mass of NH_4Cl crystals would be required to make 400. g of 15% NH_4Cl solution?

    a) 6.0×10^3 g   b) 3.75 g   c) 60 g   d) 27 g

18. The molarity of a solution containing 2.5 mol of acetic acid, $HC_2H_3O_2$ in 400. mL of solution is

  a) 0.063 M   b) 1.0 M   c) 0.103 M   d) 6.3 M

19. What is the molarity of a solution containing 0.650 mol of K_2CrO_4 in 250 mL of solution?

  a) 2.60 M   b) 0.163 M   c) 1.30 M   d) 0.326 M

20. How many moles of NaCl are present in 80 mL of 0.65 M solution?

  a) 0.052 mol   b) 123 mol   c) 8.1 mol   d) 52 mol

21. Which combination cannot form a solution?

  a) gaseous solute, gaseous solvent   b) gaseous solute, gaseous solvent
  c) gaseous solute, liquid solvent   d) solid solute, gaseous solvent

22. How would a concentration of 0.0027% KCl by mass be expressed, as part per million (ppm) of KCl?

  a) 2.7 ppm   b) 27 ppm   c) 0.27 ppm   d) 270 ppm

23. How many moles of H_2SO_4 are present in 2.20 L of 0.248 M solution?

  a) 8.87 mol   b) 0.113 mol   c) 0.546 mol   d) 113 mol

24. How many grams of K_2SO_4 are contained in 200. mL of 0.400 M K_2SO_4 solution?

  a) 0.0800 g   b) 13.9 g   c) 349 g   d) 87.2 g

25. How many grams of NaOH are required to prepare 1200. mL of 0.150 M solution?

  a) 5.00 g   b) 7.20 g   c) 222 g   d) 320 g

26. A student dissolves 196 g of H_3PO_4 in enough water to make one liter of solution. The molarity of the solution is

  a) 0.500 M   b) 2.00 M   c) 3.00 M   d) 6.00 M

27. How many milliliters of 6.0 M hydrochloric acid will contain 0.048 moles of HCl?

  a) 0.0080 mL   b) 0.29 mL   c) 8.0 mL   d) 288 mL

28. How many milliliters of 0.525 M Na_2CO_3 contain 25.0 g of Na_2CO_3?

  a) 449 mL   b) 1.39×10^3 mL   c) 13.1 mL   d) 47.6 mL

29. What is the molarity of a solution prepared by the addition of 800 mL of water to 200 mL of 3.00 M $CaCl_2$ solution? Assume volumes additive.

  a) 0.600 M   b) 0.750 M   c) 12.0 M   d) 15.0 M

30. What is the molarity of a solution prepared by mixing 300 mL of a 0.250 M solution of H_2SO_4 with 700 mL of a 6.00 M solution of H_2SO_4?

   a) 4.20 M   b) 2.14 M   c) 4.28 M   d) 6.24 M

31. If 40 mL of 0.50 M HCl are mixed with 60 mL of water, the HCl concentration of the resulting solution will be

   a) 0.20 M   b) 0.33 M   c) 1.25 M   d) 0.75 M

32. What is the molarity of a potassium bromide solution which is 22.0% KBr and has a density of 1. 18 g/mL?

   a) 0.992 M   b) 1.85 M   c) 2.18 M   d) No correct answer given.

33. 50.0 mL of 2.00 M HCl is mixed with 50.0 mL of 1.00 M NaOH and the resulting solution tested to be acidic. What is the molarity of the HCl in the final solution?

   a) 1.00 M   b) 0.0100 M   c) 0.500 M   d) 1.50 M

34. What volume of 12 M HCl is required to prepare 20 L of 0.15 M HCl?

   a) 1600 mL   b) 250 mL   c) 125 mL   d) 90 mL

35. How many milliliters of water must be added to 40.0 mL of 2.50 M KOH to prepare a solution that is 0.212 M?

   a) 3.39 mL   b) 472 mL   c) 21.2 mL   d) 432 mL

The following question set pertains to this equation:
   $2 NaOH + H_2SO_4 \rightarrow Na_2SO_4 + 2 H_2O$

36. How many grams of Na_2SO_4 can form by reacting l00 mL of 2.50M NaOH?

   a) 335 g   b) 56.8 g   c) 35.5 g   d) 17.8 g

37. How many milliliters of 4.00 M NaOH will react with 100 mL of 0.312 M H_2SO_4?

   a) 15.6 mL   b) 7.80 mL   c) 323 mL   d) 645 mL

38. How many grams of $Al(OH)_3$ can be produced from 100 mL of 0.300 M $Ba(OH)_2$?
   $3 Ba(OH)_2 + 2 AlCl_3 \rightarrow 2 Al(OH)_3 + 3 BaCl_2$

   a) 2.34 g   b) 3.51 g   c) 1.56 g   d) 0.88 g

39. What is the molality of a solution made by dissolving 40.0g of $C_2H_6O_2$ (ethylene glycol) in 400.g of water?

   a) 0.258 m   b) 0.620 m   c) 1.61 m   d) 3.88 m

40. Which of the following 1 molal solutions will have the lowest freezing temperature? (all are non-ionized).

    a) CH_3OH b) C_2H_5OH c) $C_{12}H_{22}O_{11}$
    d) None of the preceeding — all freeze at the same temperature.

41. Calculate the boiling point of an aqueous solution that freezes at –5.50°C.

    a) 101.   b) 100.55   c) 100.83   d) 100.60

42. The freezing point of an aqueous ethylene glycol, $C_2H_6O_2$, solution is –30°C. Calculate the molality of the solution

    a) 16m   b) 15m   c) 12m   d) 20m

43. Calculate the freezing point of a 20.0 m aqueous solution of ethylene glycol

    a) –15°C   b) –37.2°C   c) –28.0°C   d) –5.0°C

44. Calculate the freezing point of an aqueous solution that boils at 102.50°C

    a) –9.08°C   b) -16.5°C   c) –5.50°C   d) –25.6°C

45. Calculate the molality of a solution containing 10.0% by mass glucose, $C_6H_{12}O_6$

    a) 0.443m   b) 0.617m   c) 1.790m   d) 1.545m

46. Which of the following solutions, each made by dissolving 25 g of solute in 500 g of water, will have the lowest freezing temperature?

    a) glycerol, $C_3H_5(OH)_3$ b) glucose, $C_6H_{12}O_6$ c) ethanol, C_2H_5OH d) methanol, CH_3OH

47. Which of the following is not a colligative property?

    a) boiling point   b) vapor pressure   c) density   d) freezing point

48. If 10.0 g of a non-ionized compound dissolved in 100. g of water lowers the freezing point of the water by 0.93°C, the compound has a molar mass of

    a) 10.0 g/mol   b) 50.0 g/mol   c) 100. g/mol   d) 200. g/mol

49. An aqueous solution containing 1.50 g of compound Q in 500. g of water has a freezing point of -0.093°C. The molar mass of Q is

    a) 60.0 g/mol   b) 600. g/mol   c) 300. g/mol   d) 150. g/mol

50. If 40.0 g of ethylene glycol, $C_2H_6O_2$, are dissolved in 40.0 g of water, the freezing point of the solution will be

    a) - 16.1°C   b) - 30.0°C   c) - 33.3°C   d) No correct answer given.

51. What is the molarity of a solution with a density of 1.08 g/mL and containing 13.0% H_2SO_4?

    a) 1.52 m   b) 1.43 m   c) 1.34 m   d) 0.745 m

52. Increasing the temperature of water hinders which of the following solvents from dissolving?

    a) solid   b) liquid   c) gas   d) All of the above.

53. The solvent

    a) is always a liquid   b) must be polar to make any type of solution
    c) is present in greater amount than the solute   d) No correct answer given.

54. One liter of 2M HCl and 2 liters of 1M HCl have the same

    a) density   b) moles of solute distributed   c) concentration   d) volume

55. If NaCl is soluble in water to the extent of 36.0 g NaCl/100 g H_2O at 20°C, then a solution at 20°C containing 45 g NaCl/150 g H_2O would be

    a) dilute   b) saturated   c) supersaturated   d) unsaturated

56. The molarity of a solution containing 2.5 mol of acetic acid, $HC_2H_3O_2$, in 400. mL of solution is

    a) 0.063 M b) 1.0 M c) 0.103 M d) 6.3 M

57. What volume of 0.300 M KCl will contain 15.3 g of KCl?

    a) 1.46 L   b) 684 mL   c) 61.5 mL   d) 4.60 L

58. What mass of $BaCl_2$ will be required to prepare 200. mL of 0.150 M solution?

    a) 0.750 g   b) 156 g   c) 6.25 g   d) 31.2 g

The following two problems relate to the reaction $CaCO_3 + 2HCl \rightarrow CaCl_2 + H_2O + CO_2$

59. What volume of 6.0 M HCl will be needed to react with 0.350 mol of $CaCO_3$?

    a) 42.0 mL   b) 1.17 L   c) 117 mL   d) 583 mL

60. If 5.3 g of $CaCl_2$ were produced in the reaction, what was the molarity of the HCl used if 25 mL of it reacted with excess $CaCO_3$?

    a) 3.8 M b) 0.19 M c) 0.38 M d) 0.42 M

61. How milliliters of 6.0 M H_2SO_4 must you use to prepare 500.mL of 0.20 M sulfuric acid solution?

    a) 30   b) 17   c) 12   d) 100

62. How many milliliters of water must be added to 200. mL of 1.40 M HCl to make a solution that is 0.500 M HCl?

   a) 360 mL   b) 560 mL   c) 140 mL   d) 280 mL

63. Which of the following anions will *not* form a precipitate with silver ions, Ag^+?

   a) Cl^- b) NO_3^- c) Br^- d) CO_3^{2-}

64. Which of the following salts are considered to be soluble in water?

   a) $BaSO_4$ b) NH_4Cl c) AgI d) PbS

65. The net ionic equation for the precipitation reaction that occurs when aqueous solutions of $AgNO_3$ and K_2CrO_4 are mixed is

   a) $K^+ + NO_3^- \rightarrow KNO_3(s)$ b) $K_2 + NO_3 \rightarrow K_2NO_3(s)$
   c) $Ag^+ + CrO_4^{2-} \rightarrow AgCrO_4(s)$ d) $2\,Ag^+ + CrO_4^{2-} \rightarrow Ag_2CrO_4(S)$

66. Ocean water contains a fairly high concentration of Mg^{2+} ion. Its concentration is about 0.05 M. Ocean water therefore should not be expected to also contain a very large concentration of

   a) Cl^- b) SO_4^{2-} c) K^+ d) OH^-

67. How many grams of $AgNO_3$ are needed to prepare 1500. mL of a 0.240 M solution?

   a) 4.26 g   b) 61.2 g   c) 360. g   d) 1060 g

68. The fact that ethyl alcohol and water are very miscible is explained by the intermolecular forces known as

   a) dipole-dipole interactions   b) low molar mass of water
   c) covalent bonds   d) hydrogen bonds

69. What is the molarity of the sulfate ion in a solution prepared by dissolving 17.1 grams of aluminum sulfate, $Al_2(SO_4)_3$, in enough water to prepare 1.00 liter of solution?

   a) 0.167 M   b) 0.0500 M   c) 0.150 M   d) 0.250 M

70. Which of the following solutes, all of which are assumed to completely dissociate into ions in aqueous solution, would produce the highest concentration of phosphate ions when dissolved in water?

   a) 1.2 M $FePO_4$ b) 0.6 M $Mg_3(PO_4)_2$ c) 0.45 M $Fe_2(PO_4)_3$ d) 1.3 M Na_3PO_4

71. Consider the following two aqueous solutions at room temperature:
Solution A contains 10% NaCl by mass   Solution B contains 10% $CaCl_2$ by mass.

   All of the following statements are correct except

   a) A and B contain the same mass of solute   b) A has a higher vapor pressure than B
   c) A has a higher freezing point than B   d) A and B contain the same amount of solute

Matching

From the list of terms given choose the one that correctly identifies each phrase.

List of terms: (a) molarity; (b) supersaturated; (c) solubility; (d) miscible; (e) unsaturated; (f) molality; (g) colligative; (h) dilute; (i) saturated solution; (j) immiscible; (k) moles; (1) concentration

_____ **1.** A quantitative expression of the amount of solute in a solution.

_____ **2.** Moles solute/liter solution

_____ **3.** A term used to describe a solution containing undissolved solute in equilibrium with dissolved solute.

_____ **4.** A term used to describe a solution that is capable of dissolving more solute

_____ **5.** Molarity x Liters of solution

_____ **6.** A kind of property that depends only upon the number of solute particles in solution and not the kind of particles.

_____ **7.** The amount of one substance that will dissolve in another.

_____ **8.** Incapable of mixing and forming a solution.

_____ **9.** A term used to describe a solution that contains a relatively small amount of dissolved solute.

_____ **10.** Moles solute/kg solvent

CHAPTER 15.   Acids, Bases, and Salts

True — False (choose one)

1.   Classically, a base is a substance capable of liberating hydroxide ions, OH^-, in a water solution.

2.   The following are all properties of bases: bitter taste, soapy feeling, and the ability to change litmus from blue to red.

3.   Hydroxides of the alkali metals and the alkaline earth metals are the most common bases.

4.   In aqueous solutions it is the positive ion of the base (such as the Na^+ from $NaOH$) that is responsible for the reactions characteristic of bases.

5.   Arrhenius defined an acid as a substance that yields hydrogen ions (H^+) in aqueous solution.

6.   The Bronsted-Lowry theory defines an acid as any species that can donate a proton to another species.

7.   The Bronsted-Lowry theory defines a base as any species that can combine with a proton.

8.   The Lewis theory defines an acid as an electron-pair donor.

9. $:NH_3$ can be both an acid and a base according to the Lewis theory.

10.   A hydrated proton is a hydronium ion.

11. H_3O^+ is called an aqueous hydrogen ion.

12.   The difference between a base and its conjugate acid is a proton.

13.   In the reaction $HCl + NH_3 \rightarrow NH_4^+ + Cl^-$, the HCl and NH_3 are a conjugate acid-base pair.

14.   In the reaction $H_3PO_4 + H_2O \rightarrow H_3O^+ + H_2PO_4^-$, $H_3PO_4 - H_3O^+$ are a conjugate acid-base pair.

15.   The conjugate acid of $H_2PO_4^-$ is HPO_4^{2-}.

16.   The conjugate acid of $H_2PO_4^-$ is H_3PO_4.

17.   The conjugate base of NH_3 is NH_2^-.

18.   The sour taste common to acids is due to H_3O^+.

19. H_2O can act as both an acid and as a base.

20.   Bases react with acids to produce a salt and water.

21.   When K_2O reacts with HCl, the products are KCl and H_2O.

22.   Carbon dioxide is formed when acids react with carbonates.

23. Amphoteric hydroxides are capable of reacting as either acids or bases.

24. $Zn(OH)_2$, $Al(OH)_3$, and $NaOH$ are amphoteric hydroxides.

25. When BaO reacts with HNO_3, the products are $Ba(NO_3)_2$ and H_2O.

26. Salts are covalent compounds and generally have a high melting point and boiling point.

27. K_2S, $NaBr$, $CaSO_4$, $AlCl_3$, and $Mg(C_2H_3O_2)_2$ are all salts.

28. An electrolyte is a substance that dissolves in water and forms a solution that conducts an electric current.

29. CaO and SO_3 are electrolytes.

30. Sucrose, $C_{12}H_{22}O_{11}$, and ethyl alcohol, C_2H_5OH, are nonelectrolytes.

31. A solution made by mixing 100 mL of 0.10 M HCl and 100 mL of 0.10 M NaOH, will not conduct an electric current.

32. A solution made by mixing 100 mL of 0.10 M H_2SO_4 and 100 mL of 0.200 M NaOH, will not conduct an electric current.

33. All of the following are electrolytes: Na_2O, KCl, H_2SO_4, and $HC_2H_3O_2$.

34. The dissociation equation for $MgCl_2$ in water is

$$MgCl_2 \xrightarrow{\; H_2O \;} Mg^{2+} + 2\ Cl^-$$

35. The dissociation equation for $Cr_2(SO_4)_3$ in water is

$$Cr_2(SO_4)_3 \xrightarrow{\; H_2O \;} 3\ Cr^{3+} + 2\ SO_4^{2-}$$

36. Aqueous solutions of bases do not conduct an electric current.

37. Ions in aqueous solution are hydrated.

38. The terms strong and weak as used to describe acids refer respectively to concentrated and dilute acid solutions.

39. Most soluble salts are strong electrolytes.

40. All of the following are strong acids: HF, HCl, HBr, HI.

41. All of the following are strong electrolytes: HCl, HNO_3, NaOH, H_2CO_3.

42. A 1 M solution of $Al_2(SO_4)_3$ contains more cations than anions in solution.

43. H_2SO_3, HNO_2, $HC_2H_3O_2$, and HI are weak electrolytes.

44. Hydrogen chloride, HCl, and sodium chloride, NaCl, are ionic compounds.

45. Hydrogen chloride, HCl, and sodium chloride, NaCl, are strong electrolytes.

46. The molar concentration of all the ions in a 0.526 M $CaCl_2$ solution is 1.05 M.

47. A 1.25 M Na_2SO_4 solution contains 1.25 M Na^+ and 2.50 M SO_4^{2-}.

48. A 1.0 molal aqueous solution of KCl will freeze at about - 3.7°C.

49. A 1.0 molar solution of NaCl will freeze at a lower temperature than a 1.0 molal solution of NaCl.

50. Water can react as both an acid and a base.

51. The ionization equation for water is $H_2O + H_2O \rightleftarrows H_3O^+ + OH^-$.

52. At 25°C the H^+ ion concentration in pure water is 7×10^{-7} moles per liter.

53. At 50°C the H^+ and OH^- concentrations in pure water are equal.

54. Water ionizes to form hydrogen ions and oxide ions.

55. An acidic solution contains more H^+ ions than OH^- ions.

56. The pH is a measure of the H^+ (or H_3O^+) concentration in an aqueous solution.

57. The term pH is defined as the logarithm of the H^+ (or H_3O^+) concentration in a solution, the concentration being in moles/liter.

58. A solution with $[H^+] = 8.5 \times 10^{-2}$ has a pH between 8 and 9.

59. A pH of 7.5 is a basic solution.

60. A pH of 7.5 is an acidic solution.

61. A net ionic equation describes the actual reaction between ions of the reacting compounds.

62. Spectator ions are omitted from a net ionic equation because they do not undergo any chemical change.

63. A 0.20 M HCl solution and a 0.20 M HNO_3 solution have the same pH.

64. The pH of a solution with $[H^+] = 2.0 \times 10^{-6}$ is 5.30. (log 2 = 0.30).

65. A 0.20 M HCl and a 0.20 M $HC_2H_3O_2$ have the same pH.

66. The pH of 0.00010 M HCl is 3.0.

67. The pH of limewater is 10.5. It is a basic solution.

68. The following represents a neutralization reaction $H_3O^+ + OH^- \rightarrow 2\,H_2O$.

69. Titration is the process of measuring the volume of one reagent required to react with a measured mass or volume of another reagent.

70. For the neutralization of $HC_2H_3O_2$ with NaOH, the net ionic equation is $H^+ + OH^- \rightarrow H_2O$.

71. Carbonic acid H_2CO_3 is unstable and decomposes spontaneously into CO_2 and H_2O.

72. Neutralization is the reaction of an acid and a base to form a salt and water.

73. When 3.0 g of NaOH are added to 1000 mL of 0.10 M HCl solution, the resulting solution will be basic.

74. The number of grams of Cl^- ion in 200. mL of 0.20 M $CaCl_2$ is 1.42 g.

75. In writing net ionic equations, strong electrolytes are written in their ionic form and weak electrolytes in their molecular form.

76. For the reaction $AgNO_3$ and $BaCl_2$, the net ionic equation is $Cl_2^- + 2\,Ag^+ \rightarrow 2\,AgCl(s)$.

77. The Lewis theory defined an acid as an electron-pair donor.

78. In the reaction $HCl + NH_3 \rightarrow NH_4^+ + Cl^-$ according to the Bronsted-Lowry theory, the conjugate base of HCl is NH_4^+.

79. When an acid reacts with a carbonate, the products are a salt, water, and carbon dioxide.

80. When electrolytes dissociate, they break into individual molecules.

81. The equation for the ionization of acetic acid is $HC_2H_3O_2 + H_2O \rightleftarrows H_3O^+ + C_2H_3O_2^-$.

82. In water, zinc nitrate dissociates as shown by the equation $Zn(NO_3)_2 \xrightarrow{H_2O} Zn^{2+} + 2\,NO_3^-$.

83. The equation for the dissociation of sodium carbonate in water is $Na_2CO_3 \xrightarrow{H_2O} Na_2^{2+} + CO_3^{2-}$.

84. A solution with $[H^+] = 1.0 \times 10^{-9}\,M$ has a pH of -9.0.

85. A solution with $[H^+] = 7.8 \times 10^{-2}\,M$ has a pH between 1 and 2.

86. The reaction of an acid and a base to form a salt and water is known as neutralization.

87. For the aqueous reaction of NaOH with $FeCl_3$, the net ionic equation is $Fe^{3+} + 3\,OH^- \rightarrow Fe(OH)_3(s)$.

88. A 1.0 M $HC_2H_3O_2$ solution will freeze at a lower temperature than a 1.0 M solution of KBr.

89. Acids, bases, and salts are electrolytes.

90. Positive ions are called spectator ions.

91. Consider the reaction $2\,HCO_3^- \rightarrow H_2CO_3 + CO_3^{2-}$. In this reaction the hydrogen carbonate ion is functioning as both a Bronsted-Lowry acid and a Bronsted-Lowry base.

92. A pH meter is an instrument used to measure the acidity of a solution.

93. As the acidity of a solution increases, the pH decreases.

94. The conjugate acid of HSO_4^- is SO_4^{2-}.

95. H_3O^+ is called a hydronium ion and NH_4^+ is called a nitronium ion.

96. HNO_2, $HC_2H_3O_2$, $HClO$, and HBr are weak acids.

97. Species which can act as acids according to the Bronsted model of acids and bases are NH_4^+, CH_4, and O^{2-}.

98. The hydrogen ion concentration in a solution corresponding to a pH of 8.64 is 2.3×10^{-9}.

99. The pH of a solution of vinegar is 3.00. The concentration of OH^- in this solution is 1×10^{-3}.

100. In the Arrhenius picture of acids and bases, the reaction of an acid with a base always results in the formation of a solution which has a pH of 7.

101. The conjugate base of HPO_4^{2-} is H_3PO_4.

102. A "buffered" solution contains an insoluble solid that is amphoteric.

103. In a titration 15.0 mL of 0.100 M nitric acid neutralizes 30.0 mL of a solution of barium hydroxide. The molarity of the barium hydroxide must be 0.05 M.

104. The point in titration where the indicator changes color is called the end point of the titration.

Multiple Choice (choose the best answer)

1. In the reaction: $HCl + H_2O \rightarrow H_3O^+ + Cl^-$, which substance causes the litmus to turn red?

    a) HCl b) H_2O c) H_3O^+ d) Cl^-

2. In the reaction: $NH_3 + H_2O \rightarrow NH_4^+ + OH^-$, the water molecule serves as

    a) a proton donor b) a proton acceptor c) a weak base d) a strong base

3. In the equation: $HCO_3^- + OH^- \rightleftarrows H_2O + CO_3^{2-}$, which are the two acids according to the Bronsted-Lowry theory?

    a) HCO_3^- and CO_3^{2-} b) OH^- and H_2O c) HCO_3^- and H_2O d) OH^- and CO_3^{2-}

4. An acid-base conjugate pair for the reaction $H_3BO_3 + H_2O \rightleftarrows H_3O^+ + H_2BO_3^-$ is

    a) H_3BO_3 and H_3O^+ b) H_2O and $H_2BO_2^-$ c) H_3BO_3 and $H_2BO_3^-$ d) H_3O^+ and OH^-

5. Which is the conjugate of H_3O^+?

    a) H_2O b) OH^- c) H_2 d) H^+

6. The conjugate acid of OH^- is

     a) H_2O b) H_3O+ c) H^+ d) O^{2-}

7. Which of the following is a Lewis acid?

     a) NH_3 b) H_2O c) BF_3 d) CH_4

8. One of the products always formed when a metal reacts with an acid is

     a) a base   b) sulfur dioxide   c) a salt   d) hydrogen

9. The reaction between a proton donor and a proton acceptor is called

     a) neutralization   b) electrolysis   c) decomposition   d) single displacement

10. Which of the following is *not* a salt?

     a) $K_2Cr_2O_7$ b) $NaHCO_3$ c) $Ca(OH)_2$ d) $Na_2C_2O_4$

11. Which of the following is not a salt?

     a) $MgSO_4$ b) NCl_3 c) KNO_3 d) $CrCl_3$

12. Which equation illustrates a Lewis acid-base reaction but does *not* illustrate either Arrhenius or Bronsted-Lowry acid-base theory?

     a) $NH_3 + H_2O \rightarrow NH_4^+ + OH^-$ b) $NH_3 + BF_3 \rightarrow H_3N{:}BF_3$
     c) $NH_3 + NH_3 \rightarrow NH_4^+ + NH_2^-$ d) $NH_3 + HCl \rightarrow NH_4^+ + Cl^-$

13. Which of the following is not an electrolyte?

     a) C_2H_5OH b) $Cr_2(SO_4)_3$ c) $Ba(OH)_2$ d) $HC_2H_3O_2$

14. Which substance will not release carbon dioxide when reacted with hydrochloric acid?

     a) $MgCO_3$ b) Na_3PO_4 c) $KHCO_3$ d) Na_2CO_3

15. Which substance is an amphoteric hydroxide?

     a) $NaOH$ b) $Zn(OH)_2$ c) $Ca(OH)_2$ d) $Fe(OH)_3$

16. Which of the following is a weak electrolyte?

     a) HBr b) $Ca(NO_3)_2$ c) H_2SO_4 d) $H_2C_2O_4$

17. Which of the following is a strong electrolyte?

     a) $NaC_2H_3O_2$ b) H_2S c) HF d) HNO_2

18.   Carbonic acid, H_2CO_3, is classified as

        a) an insoluble acid   b) a strong acid   c) a weak acid   d) a nonelectrolyte

19.   Which of the following is more acidic?

        a) 1 M HCl   b) 1 M $HC_2H_3O_2$ c) 1 M H_2SO_4 d) 1 M NH_4Cl

20.   A one molal aqueous solution of which substance would have the lowest freezing point?

        a) $CaCl_2$ b) NaCl   c) $C_{12}H_{22}O_{11}$ d) CH_3OH (wood alcohol)

21.   Which solution would have the highest concentration of H^+?

        a) 1.0 M $HC_2H_3O_2$ b) 0.5 M $HC_2H_3O_2$
        c) 30 g $HC_2H_3O_2$/250 mL solution   d) 0.14 mol $HC_2H_3O_2$/140 mL solution

22.   Which solution would have the highest percent ionization?

        a) 1.0 M $HC_2H_3O_2$ b) 0.5 M $HC_2H_3O_2$
        c) 30 g $HC_2H_3O_2$/250 mL solution   d) 0.14 moles $HC_2H_3O_2$/140 mL solution

23.   Which aqueous solution would have the lowest freezing point?

        a) 1.0 molal NaCl   b) 1.0 molar NaCl
        c) 10% NaCl by weight   d) 10% sugar by weight

24.   Which 1 molal solution will have the lowest freezing point?

        a) KNO_3 b) $Mg(NO_3)_2$ c) $Al(NO_3)_3$ d) NH_4NO_3

25.   The molarity of Cl⁻ ions in a 0.20 M $CaCl_2$ solution is

        a) 0.10 M   b) 0.20 M   c) 0.30 M   d) 0.40 M

26.   The chloride ion concentration in a magnesium chloride solution is 2.0 M. The solution concentration is

        a) 2.0 M Mg^{2+} b) 2.0 M $MgCl_2$ c) 4.0 M $MgCl_2$ d) 1.0 M $MgCl_2$

27.   Choose the aqueous solution with the highest boiling point.

        a) 1 m CH_3OH b) 1 m $CaCl_2$ c) 1 m NH_4Cl d) 1 m NaCl

28.   Which 1 molal solution will have the lowest boiling point?

        a) NaCl   b) CH_3OH c) $CaCl_2$ d) $Cr(NO_3)_3$

29.   A neutral aqueous solution

        a) contains neither H^+ nor OH⁻ ions   b) always has a pH of 7
        c) contains equal concentrations of H^+ and OH⁻   d) No correct answer given.

206

30. A solution has an H^+ concentration of 2.5×10^{-4} M. The pH is

    a) 3.6 b) 4.6 c) 3.5 d) 4.4

31. The pH of an aqueous solution obtained by dissolving 20 ml of 0.1 M sodium chloride and 10 ml of 0.1 M hydrochloride acid.

    a) greater than 7 b) less than 7 c) equal to 7 d) cannot be determined

32. The pH of solution obtained by dissolving equal volume of 0.1 M NaOH and 0.2 M HCl is

    a) greater than b) equal to 7 c) less than 7 d) need to know the volume

33. A solution with a pH of 5.30 has a H^+ concentration of

    a) 5.0×10^{-5} M b) 2.0×10^{-6} M c) 7.0×10^{-6} M d) 5.0×10^{-6} M

34. A solution of which pH is most acidic?

    a) 5 b) 3 c) 9 d) 7

35. A solution of 3.65 g of HCl in 100 mL of solution would have a pH of

    a) 0 b) 1 c) 14 d) 7

36. The pH of an aqueous sodium chloride solution is

    a) greater than 7 b) equal to 7
    c) less than 7 d) depends on the molarity of the NaCl solutton

37. 14.37 mL of 0.266 M NaOH are required to titrate 10.00 mL of a hydrochloric acid solution. The molarity of the acid solution is

    a) 0.382 M b) 0.185 M c) 1.85 M d) 0.764 M

38. What volume of 0.744 M NaOH is required to titrate 10.00 mL of 0.526 M H_2SO_4?

    a) 7.07 mL b) 28.3 mL c) 26.4 mL d) 14.1 mL

39. What volume of 0.462 M NaOH is required to titrate 20.00 mL of 0.391 M HNO_3?

    a) 23.6 mL b) 16.9 mL c) 8.45 mL d) 11.8 mL

40. 16.55 mL of 0.844 M NaOH is required to titrate 10.00 mL of a hydrochloric acid solution. The molarity of the acid solution is

    a) 0.700 M b) 0.151 M c) 1.40 M d) 0.510 M

41. Which acid solution requires the smallest volume of 1.5 molar NaOH for neutralization?

    a) 25.0 mL of 1.0 M H_2SO_4 b) 25.0 mL of 1.5 M HCl
    c) 50.0 mL of 0.5 M HCl   d) 100.0 mL of 0.75 M HCl

42. What volume of concentrated (18.0 M) sulfuric acid is needed to prepare 10.0 L of 5.0 M sulfuric acid?

    a) 8.0 L   b) 2.8 L   c) 9.0 L   d) 50 L

43. $HC_2H_3O_2 + NaOH \rightarrow NaC_2H_3O_2 + H_2O$ The net ionic equation for this neutralization is

    a) $HC_2H_3O_2 + NaOH \rightarrow NaC_2H_3O_2 + H_2O$ b) $H^+ + OH^- \rightarrow H_2O$
    c) $HC_2H_3O_2 + OH^- \rightarrow C_2H_3O_2^- + H_2O$ d) No correct answer given.

44. The net ionic equation for $BaCl_2(aq) + 2\ AgNO_3(aq) \rightarrow AgCl(s) + Ba(NO_3)_2(aq)$ is

    a) $Ba^{2+} + 2\ NO_3^- \rightarrow Ba(NO_3)_2(aq)$ b) $Cl^- + Ag^+ \rightarrow AgCl(s)$
    c) $Cl_2^- + 2\ Ag^+ \rightarrow 2\ AgCl(s)$ d) No correct answer given.

45. When an ionic compound dissolves in water,

    a) the ionic solid dissociates   b) positive areas in the crystal are attracted to water's negative end
    c) ions from the crystal are hydrated   d) All of the above.

46. When aluminum becomes an ion it does so by

    a) gaining an electron   b) losing an electron
    c) gaining three electrons   d) losing three electrons

47. A strong acid

    a) ionizes ahnost completely   b) is always concentrated
    c) always tastes bitter   d) conducts electricity poorly

48. The Bronsted-Lowry definition of acids and bases centers around

    a) bases accepting electrons   b) bases accepting protons
    c) acids accepting electrons   d) acids accepting protons

49. A 0.001 M solution of H_2SO_4 could be described correctly as

    a) a concentrated solution of a strong electrolyte
    b) a dilute solution of a weak electrolyte
    c) a dilute solution of a strong electrolyte
    d) a concentrated solution of a weak electrolyte

50. When the reaction $CaO + HNO_3 \rightarrow$ is completed and balanced, a term appearing in the balanced equation is

    a) H_2 b) $2\ H_2$ c) $2\ CaNO_3$ d) H_2O

51. When the reaction $H_3PO_4 + KOH \rightarrow$ is completed and balanced a term appearing in the balanced equation is

a) H_3PO_4 b) $6 H_2O$ c) KPO_4 d) $3 KOH$

52. Which of the following is *not* a salt?

a) $K_2Cr_2O_7$ b) $NaHCO_3$ c) $Ca(OH)_2$ d) $Na_2C_2O_4$

53. Which of the following is *not* an acid?

a) H_3PO_4 b) H_2S c) H_2SO_4 d) NH_3

54. Which of the following is a weak electrolyte?

a) NH_4OH b) $Ni(NO_3)_2$ c) K_3PO_4 d) $NaBr$

55. Which of the following is a weak electrolyte?

a) $NaOH$ b) $NaCl$ c) $HC_2H_3O_2$ d) H_2SO_4

56. A solution with a pH of 5.85 has an H^+ concentration of

a) $7.1 \times 10^{-5} M$ b) $7.1 \times 10^{-6} M$ c) $3.8 \times 10^{-4} M$ d) $1.4 \times 10^{-6} M$

57. What is the pH of a $0.00015 \, M$ HCl solution?

a) 4.0   b) 2.82   c) between 3 and 4   d) No correct answer given.

58. The chloride ion concentration in 300. mL of $0.10 \, M$ $AlCl_3$ is

a) $0.30 \, M$ b) $0.10 \, M$ c) $0.030 \, M$ d) $0.090 \, M$

59. The freezing point of a 0.50 molal NaCl aqueous solution will be about

a) $-1.86°C$ b) $-0.93°C$ c) $-2.79°C$ d) No correct answer given.

60. Consider a 1.0 M solution of each of the following ionic compounds. Which will have the lowest freezing point?

a) $NaCl$ b) $BaCl_2$ c) Na_2SO_4 d) $Al(NO_3)_3$

61. The pH of a 5×10^{-8} M HNO_3 solution is

a) 6.89   b) 7.00   c) 7.30   d) 8.50

62. Which of the following equations is *not* an acid/base reaction?

a) $NaOH + HCl \rightarrow NaCl + H_2O$ b) $SO_2 + H_2O \rightarrow H_2SO_3$
c) $HCl + Zn \rightarrow ZnCl_2 + H_2$ d) $K_2O + H_2O \rightarrow 2 KOH$

63. What is the pH of a solution that is 2.5×10^{-11} M OH$^-$?

      a) 1.73   b) 3.40   c) 7.00   d) 10.60

64. What volume of 0.284 M NaOH is needed to titrate 100.00 mL of 0. 124 M HBr to the equivalence point?

      a) 35.2 mL   b) 40.8 mL   c) 43.7 mL   d) 229 mL

65. Which pair of chemical species consists of a Lewis base followed by a Lewis acid?

      a) Cl^-(aq), Ag^+(aq)   b) NH_3(g), BF_3(g)   c) SO_4^{2-}(aq), HSO_4^-(aq)   d) H^+ (aq), OH$^-$ (aq)

66. Which of the following is not a colloid?

      a) milk   b) smoke   c) cheese   d) antifreeze

Matching

From the list of terms given, choose the one that correctly identifies each phrase.

List of terms: (a) molecular equations; (b) total ionic equations; (c) net ionic equations; (d) electrolytes; (e) nonelectrolytes; (f) weak electrolytes; (g) strong electrolytes; (h) pH; (i) hydronium ion; (j) titration; (k) neutralization; (l) ionization; (m) amphoteric; (n) dissassociation; (o) inert compounds

_____ **1.** The hydrated hydrogen ion formed in water when a proton combines with a polar water molecule.

_____ **2.** Electrolytes which are essentially 100% ionized in solution.

_____ **3.** The process by which a salt, already existing as ions, separates when dissolved in water.

_____ **4.** The reaction of an acid and a base to form a salt and water.

_____ **5.** Electrolytes which are only slightly ionized in water.

_____ **6.** Substances whose solutions are nonconductors of electricity.

_____ **7.** Capable of reacting as either an acid or a base.

_____ **8.** Equations which include only those molecules or ions that have changed in the reaction.

_____ **9.** The process of measuring the volume of one reagent required to react with a measured amount or volume of another reagent.

_____ **10.** The logarithm of the reciprocal of the H^+ or H_3O^+ ion concentration in moles per liter.

CHAPTER 16. Chemical Equilibrium

True — False (choose one)

1. Many reactions do not go to completion because many reactions are reversible.

2. A reversible reaction is one in which the products formed in a chemical reaction are reacting to produce the original reactants.

3. When the equilibrium concentrations of reactants and products are reached in a chemical reaction, the chemical reaction ceases.

4. The study of reaction rates and reaction mechanisms is known as chemical kinetics.

5. A chemical equilibrium is a dynamic state in which two or more chemical reactions are going on at the same time and at the same rate but in opposite directions.

6. A saturated salt solution (undissolved salt present in the system) is in a condition of equilibrium.

7. In a chemical reaction, equilibrium is attained when the concentration of the reactants and the products are equal.

8. Although the rates of opposing reactions in a chemical equilibrium are equal, the quantities of the reactants and products may not be equal.

9. Le Chatelier's principle applies to both chemical and physical equilibria.

10. If a stress is applied to a system at equilibrium, the system will behave in such a way as to relieve that stress and restore equilibrium.

The following question set (11 – 14) pertains to this reaction:

$$4\,NH_3(g) + 5\,O_2(g) + heat \; \underset{\longleftarrow}{\overset{Pt}{\longrightarrow}} \; 4\,NO(g) + 6\,H_2O(g)$$

11. Increasing the concentration of either or both reactants forces the reaction towards more product.

12. Increasing the pressure causes the reaction to form more product.

13. Increasing the temperature will increase the amount of NO formed.

14. Increasing the amount of catalyst (Pt) will increase the amount of product.

The following question set (15-18) pertains to this reaction:

$$2\,SO_2(g) + O_2(g) \rightleftarrows 2\,SO_3(g) + 197\,kJ$$

15. When more SO_3 is added to the equilibrium system, the concentration of all the substances will increase.

16. The reaction shown is exothermic.

17. Removal of some SO_2 will decrease the concentration of O_2.

18. Decreasing the pressure will decrease the amount of O_2 in equilibrium.

19. In the reaction $H_2(g) + Br_2(g) \rightleftarrows 2\,HBr(g)$ an increase in pressure will have no effect on the equilibrium position.

20. Increasing temperature will increase the concentration of products present at equilibrium if the reaction is exothermic.

21. The presence of a catalyst will increase the yield of product present at equilibrium.

22. A catalyst lowers the energy of activation by the same amount for both the forward and reverse reactions.

23. The point of equilibrium for a reaction is the same for both the catalyzed and uncatalyzed reaction.

24. The main purpose of a catalyst in a reaction is to promote a higher yield of product at equilibrium.

25. The equilibrium constant is a numerical value that varies very little for any given equilibrium chemical reaction at a particlar temperature.

26. The magnitude of the equilibrium constant K_{eq}, is independent of temperature.

27. The magnitude of an equilibrium constant indicates how far a chemical reaction will go towards completion of the products.

28. One would expect the K_{eq} for $HCl(aq) \rightleftarrows H^+(aq)\,Cl^-(aq)$ to be a large value.

29. The magnitude of the equilibrium constant depends on the initial concentrations of the reactants.

30. The addition of a catalyst will change the value of K_{eq} for a reaction.

31. The equilibrium constant expression for $4\,NH_3(g) + 5\,O_2(g) + heat \xrightarrow{Pt} 4\,NO(g) + 6\,H_2O(g)$ is

$$K_{eq} = \frac{[NH_3]^4\,[O_2]^5}{[NO]^4\,[H_2O]^6}$$

32. The acid ionization constant expression for $HCN \rightleftarrows H^+ + CN^-$ is

$$K_a = \frac{[H^+]\,[CN^-]}{[HCN]}$$

33. Given: $K_a(HC_2H_3O_2) = 1.8 \times 10^{-5}$ and $K_a(HCHO_2) = 1.8 \times 10^{-4}$. At equal concentrations, $HCHO_2$ is a weaker acid than $HC_2H_3O_2$.

34. Diluting a solution of $HC_2H_3O_2$ causes an increase in the percent ionization of the $HC_2H_3O_2$.

35. Diluting a solution of $HC_2H_3O_2$ causes an increase in the concentration of H^+ ion in the solution.

36. The addition of the salt, $NaC_2H_3O_2$, to a solution of $HC_2H_3O_2$ results in an increase in the acidity of the solution.

37. A 0.10 M HA solution in which the HA is 1.0% ionized has a pH of 2.

38. In all water solutions the concentrations of OH^- and H^+ ions are equal.

39. In aqueous solutions at 25°C the product of the molar concentrations of H^+ and OH^- ions always equals 1×10^{-14}.

40. The ionization constant (ion product constant) for water is 1×10^{-14} at 25°C.

41. A solution of pOH 6.0 is basic.

42. A solution of pOH 9.0 is basic.

43. The $[H^+]$ in a solution of pOH = 9.0 is 1×10^{-6}.

44. A solution of $[OH^-] = 4.0 \times 10^{-8}$ has a $[H^+] = 2.5 \times 10^{-7}$.

45. A solution of $[OH^-] = 4.0 \times 10^{-8}$ has a $[H^+] = 2.5 \times 10^{-6}$.

46. The K_{sp} is determined only for slightly soluble salts.

47. The solubility product expression for $CaF_2(s) \rightleftarrows Ca^{2+} + 2 F^-$ is
$$K_{sp} = \frac{[Ca^{2+}][F^-]^2}{[CaF_2]}$$

48. $AgCl(s) \rightleftarrows Ag^+ + Cl^-$. when more $AgCl(s)$ is added to the system, the equilibrium will shift to the right.

49. The K_{sp} for Ag_2CrO_4 is 1.9×10^{-11} and the K_{sp} for $Mn(OH)_2$ is 2.0×10^{-13}. These data indicate that, in terms of moles/liter $Mn(OH)_2$ is more soluble than Ag_2CrO_4.

50. Lead(II) carbonate ($PbCO_3$), $K_{sp} = 1 \times 10^{-13}$, is more soluble in terms of moles/liter, than iron(II) carbonate, ($FeCO_3$), $K_{sp} = 2 \times 10^{-11}$.

51. When OH^- is added to a saturated solution of $Mn(OH)_2$, the basicity of the solution will increase and the solubility of $Mn(OH)_2$ will decrease.

52. A solution of sodium acetate, $NaC_2H_3O_2$, in water will have a pH greater than 7.

53. An aqueous solution of $(NH_4)_2SO_4$ will be acidic.

54. Aqueous solutions of NaC_2O_3, $NaC_2H_3O_2$, and $NaNO_3$ will be alkaline.

55. A salt derived from a strong base and a weak acid will undergo hydrolysis in an aqueous solution.

56. A solution of HCl-KCl will act as a buffer solution.

57. A solution of HCN-NaCN will act as a buffer solution.

213

58. A buffer solution reacts to maintain a fairly constant pH when small amounts of an acid or a base are added.

59. A buffer solution is a good example of the common ion effect in solution.

60. Increases in temperature always favor the forward reaction in a dynamic equilibrium.

61. For the reaction 2 CO(g) + O$_2$ (g) $\rightleftarrows$ 2CO$_2$ (g), the K$_{eq}$ = $\dfrac{[CO]^2[O_2]}{[CO_2]^2}$.

62. For an exothermic reaction, decreasing the temperature reduces the equilibrium constant.

63. If A + B $\rightleftarrows$ C + D, and the concentration of C increases, the concentration of A also will show an increase.

64. If X + Y $\rightleftarrows$ 2 Z (and all species are gaseous), the amount of Y can be increased by increasing pressure.

65. The study of reaction rates is known as chemical kinetics.

66. When the rate of the forward reaction is exactly equal to the rate of the reaction, a condition of chemical equilibrium exists.

67. The reaction CaCO$_3$(s) $\rightarrow$ CaO(s) + CO$_2$(g) will proceed to the right better in a closed container where the CO$_2$ can escape.

68. When heat is applied to a system in equilibrium, the reaction that absorbs the heat is favored.

69. The ionization constant expression for the ionization of the weak acid HCN is K_a = [H$^+$] [CN$^-$].

70. If the K_{sp} for AgI is 1.6 x 10^{-16}, and the K_{sp} for CuS is 8 x 10^{-45}, then AgI is more soluble than CuS.

71. A catalyst can lower the activation energy, thus increasing the speed of a reaction.

72. Generally, as the concentrations of the reactants increase in a chemical reaction, the speed of the reaction decreases.

73. The ion product constant for water at 25°C is 1 x10^{-14}.

74. A solution of pOH 12 has an H$^+$ concentration of 0.010 mol per liter.

75. A solution of pOH 3 will turn blue litmus to red.

76. KNO$_2$ dissolved in water will give a solution with a pH less than 7.

77. A solution made from 100 mL of 0.1 M NaOH and 100 mL of 0.2 M HC$_2$H$_3$O$_2$ will act as a buffer solution.

78. KCN will hydrolyze to give an alkaline solution.

79. The equilibrium constant expression for CH$_4$(g) + 2O$_2$(g) $\rightarrow$ CO$_2$(g) + 2H$_2$O(g) is $\dfrac{[CO_2][H_2O]}{[CH_4][O_2]}$.

80. A catalyst will increase the speed of the forward and reverse reactions equally.

81. When K_a is large (>> 1), the concentration of the reactants at equilibrium is greater than the concentration of the products.

82. If [H^+] is known, [OH^-] can be calculated from the expression for K_w.

83. If the K_a of a weak acid is 5×10^{-4}, the pH of a 0.2 M solution of that acid will be about 2.0.

84. The correct equilibrium expression for the reaction $N_2O_4(g) \rightarrow 2\ NO_2(g)$ is $K_{eq} = [N_2O_4]/[NO_2]^2$.

85. The concentrations of reactants and products vary with time in a system at chemical equilibrium.

86. A system at chemical equilibrium has a constant mass.

87. For the reaction: $2SO_2(g) + O_2(g) \rightleftarrows 2SO_3(g)$. One procedure which would result in the formation of more SO_2 would be a decrease in pressure of the system at constant temperature.

88. Consider this equilibrium reaction: $4\ NH_3(g) + 3\ O_2(g) \rightleftarrows 2\ N_2(g) + 6\ H_2O(g)$. One process which would cause this equilibrium to shift to the right would be to add a catalyst which will speed up the reaction.

89. When hydrogen and nitrogen gases react under proper conditions, ammonia gas is formed.

$$N_2(g) + 3\ H_2(g) \rightleftarrows 2NH_3(g)$$

When equilibrium is established in this situation, it is found that [N_2] = 0.02 M, [H_2] = 0.01 M, and [NH_3] = 0.10 M. Consequently, the equilibrium constant for this reaction is 5×10^5.

Multiple Choice (choose the best answer)

For the answers to the next question set (1-13), refer to the following hypothetical reactions:
(All substances are gases.)

1. $A + B \rightleftarrows 2C + 40kJ$

2. $A + 2B \rightleftarrows C_2 + 80 kJ$

3. $A + B \rightleftarrows 3C + 95 kJ$

4. $2A + B + 50 kJ \rightleftarrows 4C$

5. $2A + 3B + 35 kJ \rightleftarrows 2C_2$

1.   An endothermic reaction is

    a) (1)   b) (2)   c) (3)   d) (4)

2.   Two exothermic reactions are

    a) (4) and (5)   b) (3) and (4)   c) (1) and (5)   d) (2) and (3)

3.   The reaction which gives off the most heat per mole of product formed is

    a) (1)   b) (2)   c) (3)   d) (4)

4.   The most endothermic reaction per mole of product formed is

    a) (2)   b) (3)   c) (4)   d) (5)

5.   The reaction whose moles of product present at equilibrium would not change if the volume of the reaction vessel were decreased is

    a) (1)   b) (2)   c) (3)   d) (4)

6.   A reaction whose forward rate of reaction is increased more than the reverse rate of reaction when the temperature is increased is

    a) (3)   b) (5)   c) (2)   d) (1)

7.   A reaction whose equilibrium molar concentration of product is increased when the volume of the reaction vessel is decreased is

    a) (1)   b) (4)   c) (5)   d) all of them

8.   A reaction whose moles of product present at equilibrium are decreased when the volume of the reaction flask is increased is

    a) (1)   b) (2)   c) (3)   d) (4)

216

9. The exothermic reaction which gives off the least amount of heat per mole of product formed is

      a) (1)   b) (2)   c) (3)   d) (4)

10. A reaction whose moles of product do not change when the volume of the reaction vessel changes is

      a) (1)   b) (2)   c) (3)   d) (5)

11. A reaction whose yield of product at equilibrium is favored by both a temperature increase and decreased volume of the reaction vessel is

      a) (3)   b) (1)   c) (5)   d) (4)

12. The reactions whose yield of product at equilibrium are increased by increasing temperature are

      a) (1), (2), and (3)   b) (3) and (4)   c) (4) and (5)   d) (2) and (5)

13. The reactions whose yield of product present at equilibrium are decreased by decreasing the volume of the reaction vessel are

      a) (3) and (4)   b) (2) and (5)   c) (1), (2), and (3)   d) (4) and (5)

14. Consider the following reaction at equilibrium: $C(s) + CO_2(g) + Heat \rightleftarrows 2\ CO(g)$

The system is stressed and then returns to a new equilibrium. The stresses are

      W. Increase the amount of $C(s)$
      X. Increase the amount of $CO_2(g)$
      Y. Increase the temperature
      Z. Increase the pressure

Which of these stresses increase the amount of CO?

      a) X and Y   b) W and X   c) X   d) X, Y, and Z

15. In the stratosphere, chlorofluorocarbons, CCl_3F, produce chlorine free radical ($Cl\cdot$) that react with ozone (O_3). The following mechanism has been proposed for the destruction of the ozone layer.

      $O_3 + Cl\cdot \rightarrow ClO + O_2$ step 1
      $ClO + O \rightarrow Cl\cdot + O_2$ step 2

Identify the catalyst in this mechanism.

      a) ClO   b) Cl·   c) O   d) There is no catalyst in this reaction.

16. The following reaction is at equilibrium

      $C(s) + CO_2(g) \rightleftarrows 2\ CO(g)$

The rate of combination of C with CO_2 is 3.00 x 10-5 mole/L.s. What is the rate of decomposition of CO?

      a) 9.00×10^{-5} mole/L.s   b) 1.50×10^{-5} mole/L.s
      c) 3.00×10^{-5} mole/L.s   d) 6.00×10^{-5} mole/L.s

17. The rate of a reaction depends on

    a) the nature of the reactants   b) the temperature
    c) the presence of catalysts   d) a, b, and c are correct

18. If the reaction rate doubles for every ten degrees rise in temperature, how much faster would the reaction be at 80°C than 20°C?

    a) 32   b) 50   c) 64   d) 100

19. As a general rule, a decrease in the temperature of an aqueous saturated solution will cause the solubility of the solute to

    a) increase   b) decrease
    c) remain the same   d) have no effect on the equilibrium position

20. In the reaction $A + 2B \rightleftharpoons 2C + 3D$, the substance whose concentration would be raised to the third power in the K_{eq} expression is

    a) A   b) B   c) C   d) D

21. The equilibrium constant expression for the reaction $2A + B \rightleftharpoons 3C$ is

    a) $\dfrac{[2A][B]}{[C]}$ b) $\dfrac{[C]^3}{[A]^2[B]}$ c) $\dfrac{[A]^3[B]}{[C]^3}$ d) $\dfrac{[3C]}{[2A][B]}$

22. The equilibrium constant expression for the reaction $2A + 2B \rightleftharpoons C + 4D$ is

    a) $\dfrac{[2A][2B]}{[C][4D]}$ b) $\dfrac{[C][D]^4}{[A]^2[B]^2}$ c) $\dfrac{[A]^2[B]^2}{[C][D]^4}$ d) $\dfrac{[C][4D]}{[2A][2B]}$

23. The equilibrium constant can change significantly with a change of

    a) pressure   b) temperature   c) concentration   d) catalyst

24. Which factor will *not* increase the concentration of ammonia, NH_3?

$$N_2(g) + 3 H_2(g) \rightleftharpoons 2 NH_3(g) + 92 \text{ kJ}$$

    a) decreasing the temperature   b) increasing the concentration N_2
    c) decreasing the concentration of H_2 d) decreasing the pressure

25. At a certain temperature K_{eq} is 2.2×10^{-3} for the following reaction

$$2ICl(g) \rightleftharpoons I_2(g) + Cl_2(g)$$

Calculate K_{eq} for the reaction: $4 ICl(g) \rightleftharpoons 2 I_2(g) + 2 Cl_2(g)$

    a) 2.2×10^{-3} b) 4.8×10^{-6} c) 454   d) 4.4×10^{-3}

26. At a certain temperature K_{eq} is 2.2×10^{-3} for the following reaction

$$2ICl\,(g) \rightleftarrows I_2\,(g) + Cl_2\,(g)$$

Calculate K_{eq} for the reaction $I_2\,(g) + Cl_2\,(g) \rightleftarrows 2\ ICl(g)$

   a) -2.2×10^{-3} b) 4.8×10^{-6} c) 454   d) 4.4×10^{-3}

27. The magnitude of the equilibrium constant K_{eq} depends on

   a) initial concentration of reactants   b) presence of catalyst
   c) temperature   d) volume of the container

28. What will be the concentration of H^+ in a 0.10 M solution of HClO?

   $K_a = 3.5 \times 10^{-8}$ for HClO.

   a) $3.5 \times 10^{-7}\,M$ b) $1.9 \times 10^{-4}\,M$ c) $5.9 \times 10^{-5}\,M$ d) $5.9 \times 10^{-4}\,M$

29. What will be the concentration of H^+ in a 0.50 M solution of HClO?

   $K_a = 3.5 \times 10^{-8}$ for HClO.

   a) $1.75 \times 10^{-8}\,M$ b) $7.0 \times 10^{-8}\,M$ c) $1.3 \times 10^{-4}\,M$ d) $1.9 \times 10^{-4}\,M$

30. Which one of these solutions has the lowest pH?

   a) 0.1 M HCN   b) 0.1 M HNO_3 c) 0.1 M NaOH   d) 0.1 M H_2CO_3

31. The sodium salts of all anions below give basic solutions when dissolved in water except

   a) Cl^- b) NO^{2-} c) CN^- d) F^-

32. What is the pH of a 0.005 M $Ba(OH)_2$?

   a) 9.0   b) 10.0   c) 2.0   d) 12.0

33. A 0.20 M solution of the weak acid HA has an H^+ ion concentration of 1.4×10^{-4} mol/L. The ionization constant for this acid is

   a) 9.8×10^{-8} b) 1.8×10^{-5} c) 2.0×10^{-8} d) 7.0×10^{-4}

34. The H^+ ion concentration in a 0.10 M HCN solution is 6.3×10^{-6} M. The ionization constant for HCN is

   a) 4.0×10^{-11} b) 6.3×10^{-7} c) 6.3×10^{-5} d) 4.0×10^{-10}

35. A 0.30 M solution of formic acid, $HCHO_2$, is 2.4% ionized. The ionization constant for formic acid is

   a) 2.4×10^{-2} b) 7.2×10^{-3} c) 1.7×10^{-4} d) 1.6×10^{-5}

36. The percent ionization in a 1.0 M $HC_2H_3O_2$ solution is ($K_a = 1.8 \times 10^{-5}$)

   a) 1.34% b) 4.2% c) 0.42% d) 13.4%

37. Which statement is *not* correct?

   a) $[H^+] = \dfrac{K_w}{[OH^-]}$ b) $[OH^-][H^+] = K_w$ c) $pOH = K_w - 7$ d) $[OH^-] = \dfrac{K_w}{[H^+]}$

38. If $[OH^-] = 4.0 \times 10^{-9}$, the $[H^+] =$

   a) 2.5×10^{-5} b) 4.0×10^{-5} c) 4.0×10^{-9} d) 2.5×10^{-6}

39. An aqueous solution has a molar concentration of H^+ of 1.0×10^{-5}. The molar concentration of OH^- is

   a) 1.0×10^{-5} b) 1.0×10^{-9} c) 1.0×10^{-7} d) 1.0×10^{-10}

40. The pOH of a solution with an $[H^+] = 1 \times 10^{-5}$ is

   a) 5 b) 9 c) -9 d) -10

41. If pOH 4.0, which of the following is *not* true?

   a) $H^+ = 1.0 \times 10^{-4}$ b) pH= 10.0
   c) the solution is basic d) $[OH^-] = 1.0 \times 10^{-4}$

42. The solubility of AgBr is 7.1×10^{-7} M. The value of the solubility product, K_{sp}, is

   a) 7.1×10^{-5} b) 5.0×10^{-5} c) 5.0×10^{-13} d) 5.0×10^{-15}

43. Consider the equilibrium between AgCl and its ions in solution

$$AgCl(s) \rightleftharpoons Ag^+(aq) + Cl^-(aq).$$

Which one of the following will shift the equilibrium to the right?

   a) add $AgNO_3$ b) add NaCl c) increase temperature d) add AgCl(s)

44. The solubility product of AgI is 8.5×10^{-17}. The molar solubility of AgI in H_2O is

   a) 7.2×10^{-33} b) 9.2×10^{-8} c) 8.5×10^{-17} d) 9.2×10^{-9}

45. The solubility of CaF_2 is 2.14×10^{-4} M. The solubility product constant for CaF_2 is

   a) 3.92×10^{-11} b) 4.58×10^{-8} c) 9.80×10^{-12} d) 2.14×10^{-8}

46. The solubility of $Zn(OH)_2$ is 2.33×10^{-4} g/L. What is the solubility product constant for zinc hydroxide?

   a) 2.34×10^{-6} b) 5.48×10^{-12} c) 2.19×10^{-11} d) 5.13×10^{-17}

47. The K_{sp} expression for the slightly soluble salt PbI_2 is

a) $\dfrac{[Pb^{2+}][I^-]}{[PbI_2]}$ b) $[Pb^{2+}][I^-]$ c) $[Pb^{2+}][I^-]^2$ d) $[Pb^{2+}][2I^-]^2$

48. What will be the $[Ca^{2+}]$ when 0.020 mole of Na_2SO_4 is added to 1.0 L of saturated $CaSO_4$ solution? K_{sp} for $CaSO_4$ is 2.4×10^{-5}.

a) $4.8 \times 10^{-7}\, M$ b) $1.2 \times 10^{-3}\, M$ c) $2.4 \times 10^{-5}\, M$ d) $4.9 \times 10^{-3}\, M$

49. Which is not a valid hydrolysis equation?

a) $C_2H_3O_2^- + H_2O \rightleftarrows HC_2H_3O_2 + OH^-$

b) $Cl^- + H_2O \rightleftarrows HCl + OH^-$

c) $NH_4^+ + H_2O \rightleftarrows NH_4OH + H^+$

d) $CO_3^{2-} + H_2O \rightleftarrows HCO_3^- + OH^-$

50. An aqueous solution of which salt will be neutral?

a) $NaNO_3$ b) K_2SO_3 c) $Ca(CN)_2$ d) NH_4Cl

51. An aqueous solution of which salt will be acidic?

a) $NaNO_3$ b) K_2SO_3 c) $Ca(CN)_2$ d) NH_4Cl

52. Which pair of substances dissolved in 1.0 L of water would give an effective buffer solution?

a) 1 mole $HC_2H_3O_2$ and 0.50 mole HCl
b) 1 mole NaOH and 0.50 mole HCl
c) 1 mole $HC_2H_3O_2$ and 0.50 mole NaOH
d) 0.50 mole $HC_2H_3O_2$ and 0.5 mole NaOH

53. Which combination of acid and salt will not make a buffer solution?

a) HCN-KCN b) HCl-KCl c) $HC_2H_3O_2$-$KC_2H_3O_2$ d) HNO_2-KNO_2

54. Which of these is not necessary for equilibrium to be established?

a) All products and reactants must be present. b) The system must be closed.
c) The temperature must be increased. d) All of the above.

55. In writing the equilibrium for the melting of ice, $H_2O(s) \rightleftarrows H_2O(l)$, one would correctly choose

a) $\dfrac{[H_2O(l)]}{[H_2O(s)]}$ b) $\dfrac{1}{[H_2O(s)]}$ c) $[H_2O(l)]$ d) 1

56. What will be the H^+ concentration in a 1.0 M HCN solution? ($K_a = 4.0 \times 10^{-10}$)

a) $2.0 \times 10^{-5}\, M$ b) $1.0\, M$ c) $4.0 \times 10^{-10}\, M$ d) $2.0 \times 10^{-10}\, M$

57. What is the percent ionization of HCN in the above problem?

a) 100%   b) 2.0×10^{-8}%   c) 2.0×10^{-3}%   d) 4.0×10^{-8}%

58. If $[H^+] = 1 \times 10^{-5}$ M, which of the following is not true?

a) pH = 5   b) pOH = 9   c) $[OH^-] = 1 \times 10^{-5}$ M d) The solution is acidic.

59. The solubility product of $PbCrO_4$ is 2.8×10^{-13}. The solubility of $PbCrO_4$ is

a) 5.3×10^{-7} M b) 2.8×10^{-13} M c) 7.8×10^{-14} M d) 1.0 M

60. The solubility of AgBr is 6.3×10^{-7} M. The value of the solubility product is

a) 6.3×10^{-7} b) 4.0×10^{-13} c) 4.0×10^{-48} d) 4.0×10^{-15}

61. For the reaction $H_2(g) + I_2(g) \rightleftarrows 2\,HI(g)$, at 700 K, $K_{eq} = 56.6$. If an equilibrium mixture at 700 K was found to contain 0.55 M HI and 0.21 M H_2, the I_2 concentration must be

a) 0.046 M b) 0.025 M c) 22 M d) 0.21 M

62. In the equilibrium represented by $N_2(g) + O_2(g) \rightleftarrows 2\,NO_2(g)$ as the pressure is increased, the amount of NO_2 formed

a) increases   b) decreases   c) remains the same   d) increases and decreases irregularly

63. The solubility of $CaCO_3$ at 20°C is 0.013 g/L. What is the K_{sp} for $CaCO_3$?

a) 1.3×10^{-8} b) 1.3×10^{-4} c) 1.7×10^{-8} d) 1.3×10^{-4}

64. What will be the $[Ba^{2+}]$ when 0.010 mol of Na_2CrO_4 is added to 1.0 L of saturated $BaCrO_4$ solution? The K_{sp} for $BaCrO_4$ is 8.5×10^{-11}.

a) 8.5×10^{-11} M b) 8.5×10^{-9} M c) 9.2×10^{-6} M d) 9.2×10^{-4} M

65. A saturated solution of calcium hydroxide, $Ca(OH)_2$, in water is 0.011 M $Ca(OH)_2$. The solubility product constant for this salt is

a) 1.3×10^{-6} b) 5.3×10^{-6} c) 1.2×10^{-4} d) 2.4×10^{-4}

66. Consider the equilibrium reaction $CaCO_3(s) \rightleftarrows CaO(s) + CO_2(g)$. The reaction takes place in a closed container at a certain temperature. The most convenient way to measure the equilibrium constant for this process would be to measure

a) the pressure of the carbon dioxide gas   b) the molar concentrations of all the reactants
c) the forward and reverse rate constants   d) the density of all three species

67. The equilibrium constant for the process $Ag_2S(s) + H_2O(l) \rightleftarrows 2\,Ag^+(aq) + S^{2-}(aq)$ has a very small numerical value. This indicates that

    a) silver sulfide reacts extensively with water
    b) the solubility of silver sulfide is very small
    c) solutions can be conveniently prepared that contain large concentrations of silver and sulfide ions
    d) the process is not affected greatly by changes in temperature

68. Exactly 2.00 moles of NO and some amount of O_2 are placed in a 1-liter container at 460°C. When the reaction reaches equilibrium there are 0.00156 moles of O_2 and 0.500 mol of NO_2 present in the container. The value of the equilibrium constant is $2\,NO(g) + O_2(g) \rightleftarrows 2\,NO_2(g)$

    a) 4.42   b) 40.1   c) 71.2   d) 214

69. A catalyst that increases the rate of a reaction does so by

    a) increasing the concentrations of the initial reactants
    b) increasing the temperature
    c) decreasing the temperature
    d) decreasing the activation energy for the reaction

70. A small increase in temperature often produces a large increase in the rate of a chemical reaction because it

    a) decreases the activation energy of the reaction
    b) increases the effectiveness of the collisions between the reactant molecules
    c) decreases the number of collisions per second between the reactant molecules
    d) decreases the volume of the solution, which alters the concentrations of the reactants

71. In the potential energy diagram given, what is the value of the activation energy for the reaction $B \rightarrow A$?

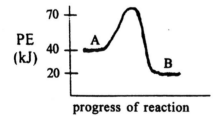

    a) 20kJ   b) 30 kJ   c) 42 kJ   d) No correct answer given.

CHAPTER 17. Oxidation–Reduction

True — False (choose one)

1. Oxidation numbers can be positive, negative, or zero.

2. A neutral atom must gain electrons in order to have a positive oxidation number.

3. The oxidation number of sulfur in S_8 is the same as the oxidation number of phosphorus in P_4.

4. Oxygen has the same oxidation number in ozone (O_3) and oxygen (O_2).

5. The oxidation number of oxygen in H_2O is the same as the oxidation number of oxygen in H_2O_2.

6. When two atoms share a pair of electrons, the atom with the higher electronegativity will have the positive oxidation number.

7. Hydrogen may have an oxidation number of - 1, 0, or + 1.

8. The algebraic sum of the oxidation numbers of the elements in a polyatomic ion must equal zero.

9. An increase in oxidation number corresponds to a loss of electrons.

10. The oxidation number of aluminum in Al^{3+} is + 3.

11. The oxidation number of iodine in KIO_3 is + 3.

12. The oxidation number of carbon in $NaHCO_3$ is + 5.

13. The oxidation number of uranium in UO_2^{2+} is + 6.

14. The oxidation number of chromium in $Cr_2O_7^{2-}$ is + 6.

15. The nitrogen atom in N_2O_3 and NO_2 have the same oxidation number.

16. An element gaining electrons will become less positive or more negative.

17. Oxidation and reduction always occur simultaneously in a chemical reaction.

18. The oxidating agent in a chemical reaction increases its oxidation number.

19. In some reactions both the oxidizing agent and the reducing agent gain electrons.

20. In the reaction $2\ F_2 + O_2 \rightarrow 2\ OF_2$ the oxygen oxidizes the fluorine.

21. In a redox reaction the reducing agent is oxidized and the oxidizing agent is reduced.

22. A reducing agent is a substance that accepts electrons in a chemical reaction.

23. An oxidizing agent is a substance that accepts electrons in a chemical reaction.

24. In the reaction: $MnO_2 + 4HCl \rightarrow MnCl_2 + Cl_2 + 2 H_2O$, the reducing agent is MnO_2.

25. $Zn \rightarrow Zn^{2+} + 2 e-$ is an oxidation half-reaction.

26. $ClO_4^- + 4 H^+ \rightarrow ClO_2^- + 2 H_2O$ is a balanced reduction half-reaction.

27. Five electrons are needed on the left side of the equation to balance this half-reaction:

$$MnO_4^- + 8 H^+ \rightarrow Mn^{2+} + 4 H_2O$$

28. When this half-reaction is balanced, it will contain four electrons on the right side of the equation:

$$BrO_2^- + H_2O \rightarrow BrO_4^- + H^+$$

29. In a voltaic cell a chemical change is used to produce electrical energy.

30. In a voltaic cell electrical energy is used to produce a chemical change.

31. In a galvanic cell, electrons flow from the anode to the cathode.

32. A spontaneous oxidation-reduction reaction takes place in a voltaic cell.

33. An electrolytic cell does not require a battery.

34. The salt bridge completes the circuit in a galvanic cell.

35. In an electrolytic cell, electrical energy is used to produce a chemical change.

36. In the electrolysis of sodium chloride brine, Cl^- is oxidized to Cl_2 at the anode.

37. A charged storage battery acts as a voltaic cell when it furnishes energy to operate a radio.

38. The electrolysis of molten NaCl and aqueous NaCl yields the same products.

39. An important electrochemical application is electroplating of metals.

40. Oxidation occurs at the anode in a voltaic cell and at the cathode in a galvanic cell.

41. The most common oxidation number of the alkali metals in compounds is + 1.

42. In general, substances that are easily oxidized have high electronegativities.

43. When $O_2 \rightarrow 2O^{2-}$, electrons are released.

44. H_2SO_3, SO_2 and SO_3^{2-} all contain sulfur in the same oxidation state.

45. In a battery, the cathode is the terminal to which positive ions are attracted.

46. Oxidation and reduction occur simultaneously in a chemical reaction; one cannot take place without the other.

47. The negative electrode is called the cathode.

48. The cathode is the electrode at which oxidation takes place.

49. Metallic zinc will react with hydrochloric acid.

50. In electroplating, the piece to be electroplated with a metal is attached to the cathode.

51. Potassium is a better reducing agent than sodium.

52. $2Ag(s) + 2HCl(aq) \rightarrow 2AgCl(s) + H_2(g)$

53. In a voltaic cell, reduction occurs at the cathode.

54. The oxidation number of P in $Mg_2P_2O_7$ is +7.

55. $CrO_4^{2-} + 4 H_2O + 2 e- \rightarrow Cr(OH)_3 + 5 OH^-$ is a balanced reduction half-reaction.

56. It is very common in electroplating to deposit a metal from the solution of some salt or ion of the metal. This deposition occurs at the anode of the cell where electrons are lost by the metal.

57. In the oxidation-reduction reaction

$$Sn^{4+} + 2 Fe^{2+} \rightarrow 2 Fe^{3+} + Sn^{2+}$$

Sn^{4+} is the reducing agent and Fe^{3+} is the oxidizing agent.

58. In the half-reaction whereby $S_2O_3^{2-}$ chemically changes into $S_4O_6^{2-}$, the number of electrons involved in the process for each sulfur atom reacting is 1/2.

59. It would be safe to store an acidic solution (H^+) in a container made of aluminum metal.

60. In oxidation-reduction reactions, it is important to balance atoms, electrical charge, and heat energy on both sides of the equation.

61. Effective oxidizing agents are substances which might likely have high electronegativities.

Multiple Choice (choose the best answer)

1. In which substance is the oxidation number of oxygen different from the oxidation number of oxygen in the other three substances?

    a) OH^- b) H_2O c) H_2O_2 d) O^{2-}

2. In which compound is the oxidation number of hydrogen negative?

    a) NaH b) H_2 c) H_2O d) CH_4

3. The oxidation number of chlorine in Cl_2O_7 is the same as the oxidation number of chlorine in

    a) NaCl   b) $KClO_4$ c) $HClO_3$ d) Cl_2

4. Which element in $Ba(ClO_3)_2 \rightarrow H_2O$ has the same oxidation number as chlorine in HClO?

    a) Ba   b) Cl   c) O   d) H

5. In which pair of compounds is the oxidation number of nitrogen and chlorine different?

    a) N_2O_3 and $Ba(ClO_2)_2$ b) N_2O_5 and $KClO_3$
    c) N_2O and HClO   d) NO_2 and $KClO_4$

6. The oxidation number of iodine in KIO_4 is

    a) - 1   b) + 1   c) + 3   d) +7

7. The oxidation number of boron in BO_3^{3-} is

    a) +6   b) + 5   c) + 3   d) - 1

8. An example of an oxidation-reduction reaction is

    a) $BaCl_2 - 2H_2O \rightarrow 2 BaCl_2 + H_2O$ b) $2 Al + 3 Cl_2 \rightarrow 2 AlCl_3$
    c) $KOH + HI \rightarrow KI + H_2O$ d) $Fe^{2+} + S^{2-} \rightarrow FeS$

9. The principal characteristic of all redox reactions is

    a) the formation of water
    b) the transfer of electrons
    c) the formation of a gas or precipitate
    d) an irreversible reaction

10. Which one of the following is the most easily oxidized?

    a) Ca   b) Ba   c) Cu   d) Cr

11. The nitrogen with a +3 oxidation state is

    a) N_2O b) N_2O_3 c) N_2O_4 d) NO

12. An oxidizing agent is a substance that

    a) shares electrons   b) gains electrons
    c) loses electrons   d) forms ionic bonds

13. In the reaction, $4 Br_2 + H_2S + 4 H_2O \rightarrow H_2SO_4 + 8 HBr$, the element oxidized is

    a) Br   b) H   c) S   d) O

14. In the reaction, $3 CuO + 2 NH_3 \rightarrow N_2 + 3 H_2O + 3 Cu$, the reducing agent is

    a) CuO b) NH_3 c) N_2 d) H_2O

15. In the equation $Mg + H_2SO_4 \rightarrow MgSO_4 + H_2$, the element reduced is

    a) Mg b) H c) S d) O

16. In the unbalanced half-reaction $SO_4^{2-} \rightarrow S^{2-}$, the sulfur in SO_4^{2-}

    a) gains 6 e-   b) loses 6 e-   c) gains 8 e-   d) loses 8-

17. When the following equation, $As_2O_3 + Cl_2 + H_2O \rightarrow H_3AsO_4 + HCl$, is balanced, a term in the balanced equation is

    a) $2 As_2O_3$ b) $3 Cl_2$ c) $3 H_3AsO_4$ d) $5 H_2O$

18. When the following equation, $Cl_2 + KOH \rightarrow KClO_3 + KCl + H_2O$, is balanced, a term in the balanced equation is

    a) $3 Cl_2$ b) $4 KOH$ c) $3 KCl$ d) $5 H_2O$

19. When the equation, $Cu + H^+ + NO_3^- \rightarrow NO + H_2O + Cu^{2+}$, is balanced, a term in the balanced equation is

    a) $2 Cu$ b) $3 NO_3^-$ c) $3 Cu^{2+}$ d) $7 H_2O$

20. If the following equation, $NO_3^- + I^- \rightarrow NO + I_2$, is balanced by *the ion-electron method in acid solution,* a correct half-reaction is

    a) $I^- \rightarrow I_2 + e^-$ b) $NO_3^- + 4H^+ + 3 e^- \rightarrow NO + 2 H_2O$
    c) $N^{6+} + 4 e^- \rightarrow N^{2+}$ d) $NO_3^- + 3 e^- \rightarrow NO$

21. If the following equation, $NO_2^- + Al \rightarrow NH_3 + AlO_2^-$, is balanced by the *ion-electron method in basic solution*, a correct half-reaction is

    a) $Al + 4 OH^- \rightarrow AlO_2^- + 2 H_2O + 3 e^-$ b) $NO_2^- + 7 H^+ + 6 e^- \rightarrow NH_3 + 2 H_2O$
    c) $Al \rightarrow Al^{3+} + 3 e-$ d) $NO_2^- + 5 H_2O + 8e^- \rightarrow NH_3 + 7 OH^-$

22. When the following equation is balanced, $MnO_4^- + AsO_3^{3-} + H^+ \rightarrow Mn^{2+} + AsO_4^{3-} + H_2O$, a term in the balanced equation is

    a) MnO_4^- b) $5 AsO_4^{3-}$ c) $4 AsO_3^{3-}$ d) $4 H_2O$

23. If the following equation, $BiO_3^- + Mn^{2+} \rightarrow Bi^{3+} + MnO_4^-$, is balanced by the *ion-electron method in acid solution*, a correct term in the balanced equation is

    a) $3 BiO_3^-$ b) $4 Mn^{2+}$ c) $6 H^+$ d) $7 H_2O$

24. If the following equation, $Cr^{3+} + HClO_4 \rightarrow Cr_2O_7^{2-} + Cl^-$, is balanced by the *ion-electron method in acid solution*, a correct term in the balanced equation is

    a) $6\,Cr^{3+}$ b) $3\,HClO_4$ c) $2\,Cr_2O_7^{2-}$ d) $18\,H^+$

25. If the short version of the Activity Series of Metals is K, Ca, Mg, Al, Zn, Fe, H_2, Cu, Ag, which of the following pairs will not react in a water solution?

    a) $Zn, CuSO_4$ b) $Cu, Al_2(SO_4)_3$ c) $Fe, AgNO_3$ d) $Ca, Al_2(SO_4)_3$

26. If a short version of the Activity Series of Metals is K, Ca, Mg, Al, Zn, Fe, H_2, Cu, Ag, which of the following is a correct equation?

    a) $Fe + 3\,HCl \rightarrow FeCl_3 + 3\,H$ b) $2\,AgCl + H_2SO_4 \rightarrow Ag_2SO_4 + 2\,HCl$
    c) $Cu + ZnSO_4 \rightarrow Zn + CuSO_4$ d) $H_2 + 2\,AgNO_3 \rightarrow 2\,HNO_3 + 2\,Ag$

27. When an automobile lead storage battery is discharging, which of these is a proper cathode reaction?

    a) $PbO_2 + 4\,H^+ + 2\,e^- \rightarrow Pb^{2+} + 2\,H_2O$ b) $Pb^{2+} + SO_4^{2-} \rightarrow PbSO_4$
    c) $Pb \rightarrow Pb^{2+} + 2\,e^-$ d) $Pb + PbO_2 + 2\,H_2SO_4 \rightarrow 2\,PbSO_4 + 2\,H_2O$

28. Three metals, A, B, and C, are listed in order of chemical activity. Which of the following statements is true?

    a) atoms of B can reduce ions of A   b) atoms of B can reduce ions of C
    c) atoms of B can reduce atoms of A   d) atoms of B can reduce atoms of C

29. When the following equation is balanced in an alkaline solution, $MnO_4^- + Cl^- \rightarrow MnO_2 + Cl_2$, a term in the balanced equation is

    a) MnO_4^- b) $4\,Cl^-$ c) $2\,H_2O$ d) $8\,OH^-$

30. When the following equation is balanced in an alkaline solution a term in the balanced equation is
    $$PbO_2 + Sb \rightarrow PbO + SbO_2^-$$

    a) $2\,PbO_2$ b) $2\,Sb$ c) $2\,OH^-$ d) $2\,H_2O$

31. How many grams of sulfur can be obtained by reacting H_2S with 50.0 g of HNO_3?
    $$3\,H_2S + 2\,HNO_3 \rightarrow 3S + 2\,NO + H_2O$$

    a) 25.5 g   b) 38.2 g   c) 47.1 g   d) 70.6 g

32. How many liters of NO gas at STP can be produced when 2.52 mol of Cu are reacted?
    $$Cu + NO_3^- \rightarrow Cu^{2+} + NO \quad \text{(acid solution)}$$

    a) 56.4 L   b) 84.8 L   c) 37.6 L   d) No correct answer given.

33. In the reaction $Na + H_2O \rightarrow NaOH + H_2$, which of these is true?

    a) Sodium is oxidized   b) Water is oxidized
    c) Hydrogen is oxidized   d) Oxygen is oxidized

34. In an electrolytic cell, KI is electrolyzed and at the anode, I- is oxidized. Which of these represents the anode reaction?

    a) $I^- + e^- \rightarrow I$ b) $2I^- \rightarrow I_2 + 2e^-$ c) $2I^- + 2e^- \rightarrow I_2$ d) $I_2 \rightarrow 2e^- + 2I^-$

35. In the reaction $H_2S + 4Br_2 + 4H_2O \rightarrow H_2SO_4 + 8HBr$, the oxidizing agent is

    a) H_2S b) Br_2 c) H_2O d) H_2SO_4

36. In the reaction $VO_3^- + Fe^{2+} + 4H^+ \rightarrow VO^{2+} + Fe^{3+} + 2H_2O$ the element reduced is

    a) V   b) Fe   c) O   d) H

37. In the partially balanced redox equation $3Cu + HNO_3 \rightarrow 3Cu(NO_3)_2 + 2NO + H_2O$ the coefficient needed to balance H_2O is

    a) 8   b) 6   c) 4   d) 2

38. Which reaction does not involve oxidation-reduction?

    a) Burning sodium in chlorine   b) Chemical union of Fe and S
    c) Decomposition of $KClO_3$ d) Neutralization of NaOH with H_2SO_4

39. How many moles of Fe^{2+} can be oxidized to Fe^{3+} by 2.50 mol of Cl_2 according to the following equation?
    $$Fe^{2+} + Cl_2 \rightarrow Fe^{3+} + Cl^-$$

    a) 2.50 mol   b) 5.00 mol   c) 1.00 mol   d) 22.4 mol

40. How many grams of sulfur can be produced from 100. mL of 6.00 M HNO_3?
    $$HNO_3 + H_2S \rightarrow S + NO + H_2O$$

    a) 28.9 g   b) 19.3 g   c) 32.1 g   d) 289 g

41. If the smallest whole-number coefficients are used to balance this oxidation-reduction reaction,
    $$Na_2CrO_4 + CO_2 + KNO_2 \rightarrow Cr_2O_3 + Na_2CO_3 + KNO_3$$
    the sum of the coefficients in the balanced equation will be

    a) 13   b) 14   c) 16   d) 19

42. Consider the unbalanced equation
    $$Fe^{2+} + MnO_4^- + H^+ \rightarrow Mn^{2+} + Fe^{3+} + H_2O$$
    When correctly balanced with the smallest whole number coefficients, the sum of the coefficients in the answer is

    a) 12   b) 18   c) 22   d) 24

43. In a voltaic cell, oxidation occurs at the

    a) anode   b) cathode   c) salt bridge
    d) electrode at which electrons enter from the outside

44. In the chemical decomposition of hydrogen peroxide

$$2 H_2O_2 \rightarrow 2 H_2O + O_2$$

the role of the oxygen in the hydrogen peroxide is that of

    a) oxidizing agent
    b) reducing agent
    c) both oxidizing and reducing agent
    d) neither the oxidizing nor the reducing agent

45. All of the following are true for a galvanic cell except

    a) oxidation occurs at the anode
    b) cations are attracted to the cathode
    c) a non-spontaneous reaction takes place in the cell
    d) the salt bridge contains electrolytes to maintain neutrality in the cell

46. All of the following involve a redox reaction except:

    a) electrolysis of HCl   b) corrosion of Fe   c) electroplating of Ni   d) ionization of HCl

47. Consider the following reaction taking place in a galvanic cell:

$$Sn^{2+}(aq) + Cu(s) \rightarrow Sn(s) + Cu^{2+}(aq)$$

The reaction occurring at the anode is:

    a) $Sn^{2+}(aq) + 2e \rightarrow Sn(s)$
    b) $Sn(s) \rightarrow Sn^{2+}(aq) + 2e$
    c) $Cu(s) \rightarrow Cu^{2+}(aq) + 2e$
    d) $Cu^{2+} + 2e \rightarrow Cu(s)$

Balancing Oxidation-Reduction Equations (**Balance each of the following equations**)

1. $MnSO_4 + PbO_2 + H_2SO_4 \rightarrow HMnO_4 + PbSO_4 + H_2O$

2. $Cr_2O_7^{2-} + Cl^- \rightarrow Cr^{3+} + Cl_2$ (acidic solution)

3. $MnO_4^- + AsO_3^{3-} \rightarrow Mn^{2+} + AsO_4^{3-}$ (acidic solution)

4. $Zn + NO_3^- \rightarrow Zn(OH)_4^{2-} + NH_3$ (basic solution)

5. $KOH + Cl_2 \rightarrow HCl + KClO + H_2O$

6. $As + ClO_3^- \rightarrow H_3AsO_3 + HClO$ (acidic solution)

7. $H_2O_2 + Cl_2O_7 \rightarrow ClO_2^- + O_2$ (basic solution)

8. $S_2O_3^{2-} + HNO_3 \rightarrow S_4O_6^{2-}$

9. $MnO_4^- + C_2O_4^{2-} \rightarrow Mn^{2+} + CO_2$ (acidic)

10. $HNO_2 + H_2S \rightarrow S + NO + H_2O$

CHAPTER 18. Nuclear Chemistry

True — False (choose one)

1.   Radioactivity was discovered by Wilhelm Roentgen.

2.   The atomic number of $^{66}_{30}X$ is 36.

3.   An atom of $^{226}_{88}Ra$ has a total of 314 nucleons.

4.   Radioactivity is the spontaneous emission of radiation from the nucleus of an atom.

5.   The radioactive process is not affected by changes in physical state or temperature.

6.   The nature of the radioactivity exhibited by a radioactive element depends on the form or compound in which the element exists.

7.   The principal emissions from the nucleus of disintegrating natural radionuclides are known as alpha rays or particles, beta rays or particles, and gamma rays.

8.   All atoms with atomic numbers greater than 83 are radioactive.

9.   In two half-life periods of time a radioactive substance will decay completely (100%) into another element.

10.   Radionuclides of the same element have the same half-lives.

11.   All particles and rays emanate from the nucleus in radioactive changes.

12.   A positron has a positive charge of + 1.

13.   The half-life of radionuclides can vary from millionths of a second to billions of years.

14.   The longer the half-life of a radionuclide, the faster it decays.

15.   If the half-life of Ra-226 is 1620 years, the amount remaining from 1000 g of Ra-226 at the end of 4860 years is 125 g.

16.   As the temperature of a radionuclide increases, its half-life decreases.

17.   The following symbols all represent an alpha particle: α, He^{2+}, $^{4}_{2}He$.

18.   A beta particle is written as β or $^{0}_{-1}e$.

19.   A balanced nuclear reaction has the same number and kinds of atoms on both sides of the equation.

20.   When alpha, beta, or gamma radiation is emitted from a nucleus, the residue remaining is a different element.

21. In order for an element to change into another element (transmutation) the number of protons in the nucleus must change.

22. Alpha disintegration of a radionuclide results in the formation of a new element with an atomic number four less than the starting element.

23. The half-life of Ra-226 is 1620 years.

24. When an atom emits a beta particle as a result of nuclear decay, the number of protons in the atom increases.

25. The loss of one alpha and two beta particles from $^{211}_{84}$Po leaves $^{207}_{84}$Po.

26. The disintegration of $^{247}_{96}$Cm to $^{235}_{92}$U involves the loss of four alpha and four beta particles from the nucleus.

27. The disintegration of $^{247}_{96}$Cm to $^{235}_{92}$U involves the loss of three alpha and two beta particles from the nucleus.

28. The gamma ray has the greatest penetrating ability of the three principal emissions from the nucleus.

29. Radioactivity is due to an unstable ratio of neutrons to protons in an atom.

30. The three naturally occurring radioactive disintegration series all end with a stable nuclide of lead.

31. The stable end-product of the natural disintegration series of U-238 is Pb-206.

32. A transmutation reaction must involve a change in the number of protons in the nucleus of an atom.

33. The first artificial transmutation was accomplished by Ernest Rutherford in which he converted $^{14}_{7}$N to $^{17}_{8}$O.

34. The bombardment of stable nuclei with atomic particles to form unstable nuclei which are radioactive is called induced or artificial radioactivity.

35. Artificial radionuclides do not have specific half-lives like natural radionuclides.

36. The symbol for the missing particle in the equation $^{27}_{13}$Al + $^{4}_{2}$He → ? + $^{1}_{0}$n is $^{30}_{15}$p.

37. The unit of measurement of radioactivity is the curie, which corresponds to the amount of radiation produced by 1 gram of pure radium.

38. A Geiger counter operates on the basis that radiations from a radioactive source are able to ionize molecules, forming ions and creating a flow of a detectable electric current.

39. Two or more neutrons are released during a nuclear fission reaction.

40. In nuclear fission a heavy nuclide splits into two or more smaller nuclides.

41. Most nuclides formed as a result of nuclear fission are radioactive.

42. The neutrons released during nuclear fission are necessary to sustain a nuclear chain reaction.

43. The major components of a nuclear power plant are nuclear fuel, a central system to regulate the heat generated, and a cooling system which returns the heat from the nuclear reactor.

44. A nuclear reactor is a special furnace that produces energy by a controlled nuclear chain reaction.

45. Nuclear fusion is the process of uniting the nuclei of two light elements to form one heavier nucleus.

46. Helium may be produced by the fusion of hydrogen isotopes.

47. The energy from nuclear fusion reactions is derived by the conversion of mass into energy.

48. The difference in mass between the mass of a nucleus and the total mass of the protons and neutrons in the nucleus is called the mass defect.

49. The binding energy of a nucleus is the energy equivalent of the mass defect of the nucleus.

50. The smaller the mass defect of a nucleus the greater the binding energy.

51. The equivalence between mass and energy is expressed by $E = 1/2mv_2$.

52. The elements following uranium are called transuranium elements.

53. Almost all the known transuranium elements are radioactive.

54. Alpha, beta, and gamma radiations are classified as ionizing radiations and can damage or kill living cells.

55. Strontium-90 is a hazardous radionuclide because it deposits in the bones like calcium and its radiation may cause leukemia and bone cancer.

56. Nuclear radiation may damage the structure of the genes which can then have an effect on the genetic traits of future generations.

57. Radioactive tracers are compounds containing a radionuclide whose reactions in a chemical or biological system can be traced or followed using radiation detection instruments.

58. Radiocarbon age dating is based on a diminishing ratio of C-14 to C-12 in a previously living species.

59. Radioactivity was first discovered by Antoine Henri Becquerel.

60. A thin sheet of paper will normally block beta radiation.

61. When a nucleus emits an alpha particle or a beta particle, it always changes into a nuclide of a different element.

62. When an atom loses a beta particle from its nucleus, a different element is formed, having essentially the same mass and atomic number one greater than the starting element.

63. A radioactive disintegration series shows the succession of alpha and beta emissions by which naturally occurring radioactive elements decay to reach stability.

64. The energy from nuclear fission can be harnessed to produce steam which can drive turbines and produce electricity.

65. The high energy released from nuclear fusion gives other nuclei enough kinetic energy to sustain a chain reaction.

66. Cancers are often treated by gamma radiation from CO–60, which destroys the rapidly growing cancer cells.

67. Protracted exposure to low levels of any form of ionizing radiation can weaken the body and lead to the onset of malignant tumors.

68. A positron has the mass of an electron and the electrical charge of a proton.

69. Transmutation is the changing of a nuclide of one element to a nuclide of another element.

70. Rem is the unit of radiation that relates to the biological effect of absorbed radioactive radiation.

71. A Geiger counter measures ionizing radiation.

72. An alpha particle is a negatively charged particle with a mass number equivalent to that of a proton.

73. Cesium-137 spontaneously decays to emit beta particles and to form barium-137. The half-life of the cesium isotope is about 30 years. It will take approximately 150 years before 97% of the cesium in a particular sample has decomposed.

74. An isotope of uranium undergoes a fission reaction when bombarded with neutrons. The mass number of the uranium isotope used must have been 238. $_{0}^{1}\text{n} + {}_{92}\text{U} \rightarrow {}_{56}\text{Ba}^{141} + {}_{36}\text{Kr}^{92} + 3\,{}_{0}^{1}\text{n}$.

75. Atoms of all elements have at least one neutron in the nucleus.

76. These nuclear particles are listed in order of increasing ability to penetrate matter: α, β, γ.

77. After the duration of three half-lives, 1/6 of any radioactive sample will still be radioactive.

78. Half-life and the stability of a nuclide are indirectly related.

79. A critical mass is required to sustain a chain reaction in a nuclear fission.

80. Gamma radiation is affected by electromagnetic field.

Multiple Choice (choose the best answer)

1. Radioactivity was discovered by

    a) Becquerel   b) Curie   c) Rutherford   d) Roentgen

2. Radioactive changes differ from ordinary chemical change in that radioactive changes

    a) are explosive   b) absorb energy   c) involve changes in the nucleus   d) release energy

3. The particle among the following having the least mass is the

    a) electron   b) proton   c) neutron   d) a particle

4. The symbol $_1^2 H$ represents a

    a) neutron   b) deuteron   c) triton   d) hydrogen atom

5. The isotopes U-235 and U-238 differ in

    a) atomic number   b) mass number
    c) number of valence electrons   d) number of protons in the nucleus

6. Which pair of particles are isotopes?

    a) $_1^3 H$ and $_1^2 D$ b) $_{93}^{239} Np$ and $_{94}^{239} Pu$ c) $_{-1}^{0} e$ and $_{+1}^{0} e$ d) $_2^4 He$ and $_2^4 He^{2+}$

7. The half-life of Sn-110 is 4 hours. If you had 40 grams of this nuclide, how much would you have left 8 hours later?

    a) 5 grams   b) none   c) 20 g   d) 10 g

8. If $_{91}^{234} Pa$ loses a beta particle, the resulting isotope is

    a) $_{89}^{230} Ac$ b) $_{90}^{234} Th$ c) $_{92}^{234} U$ d) $_{90}^{233} Th$

9. A radionuclide has a half-life of 5.0 years. After 25 years of decay, what fraction of the original nuclide is left?

    a) 1/8   b) 1/5   c) 1/25   d) 1/32

10. In the equation $_7^{14} N + _0^1 n \rightarrow C + _1^1 H$ the mass number of the carbon nuclide formed is

    a) 6   b) 15   c) 14   d) 12

11. The particle or ray having the greatest penetrating ability is

    a) α b) β c) γ d) positron

12. If $_{84}^{222} Rn$ loses an alpha particle, the resulting nuclide

    a) $_{82}^{218} Pb$ b) $_{85}^{222} At$ c) $_{87}^{222} Fr$ d) $_{84}^{218} Po$

239

13. In the equation, $^{53}_{24}Cr + ? \rightarrow \,^{53}_{23}V + \,^{1}_{1}H$, the missing bombarding particle is

   a) $^{0}_{-1}e$ b) $^{1}_{0}n$ c) $^{2}_{1}H$ d) $^{4}_{2}He$

14. In the equation, $^{253}_{99}Es + \,^{4}_{2}He \rightarrow ? + \,^{1}_{0}n$, the missing nucletide is

   a) $^{257}_{100}Fm$ b) $^{256}_{101}Md$ c) $^{249}_{97}Bk$ d) $^{257}_{101}Md$

15. $^{24}_{11}Na$ decays with a half-life of 15 hours. If you started with 20. g of this isotope, how much would you have left after 45 hours?

   a) 20. g   b) 5.0 g   c) 2.5 g   d) 1.3 g

16. $^{234}_{92}U$ successively emits $\alpha, \alpha, \alpha, \alpha, \alpha, \beta$. At that point the nuclide has become

   a) $^{214}_{83}Bi$ b) $^{218}_{85}At$ c) $^{214}_{81}Ti$ d) $^{213}_{83}Bi$

17. In the equation $^{239}_{93}Np \rightarrow \,^{239}_{94}Pu + X$ the symbol X is

   a) an electron   b) a proton   c) a neutron   d) a positron

18. Which of the following represents the reaction?

   $^{40}_{18}Ar + \,^{1}_{1}H \rightarrow \,^{40}_{19}K + \,^{1}_{0}n$

   a) $^{40}_{18}Ar(^{1}_{1}H, \,^{1}_{0}n)^{40}_{19}K$ b) $^{40}_{19}Ar(^{1}_{0}n, \,^{1}_{1}H)^{40}_{19}K$

   c) $^{40}_{19}K(^{1}_{1}H, \,^{1}_{0}n)^{40}_{18}Ar$ d) $^{40}_{19}K(^{1}_{0}n, \,^{1}_{1}H)^{40}_{18}Ar$

19. A bismuth isotope, $^{214}_{83}Bi$, can be formed by either an α decay or β decay. Two particles that can do this are

   a) $^{214}_{84}Po, \,^{218}_{85}At$ b) $^{214}_{82}Pb, \,^{210}_{81}Tl$ c) $^{218}_{84}Po, \,^{210}_{83}Bi$ d) $^{214}_{82}Pb, \,^{218}_{85}At$

20. Which of the following is *not* a unit of radiation?

   a) Curie   b) Roentgen   c) Ram   d) Gray

21. Which of the following is *not* a characteristic of nuclear fission?

   a) Upon absorption of a neutron, a heavy nucleus splits into two or more smaller nuclei.
   b) One neutron is produced from the fission of each atom.
   c) Large quantities of energy are produced.
   d) Most nuclei formed are radioactive.

22. If U-235 is bombarded by a neutron, the atom can split into

   a) Sr and Pb   b) Cd and Kr   c) Ba and Xe   d) Ba and Kr

23. Given the equation: $_{92}^{235}U + _0^1n \rightarrow {}_{38}^{90}Sr + {}_{54}^{140}Xe + Y\ _0^1n$. Y represents

       a) 0   b) 2   c) 5   d) 6

24. The radiation called alpha particles is composed of

       a) gamma rays   b) positrons   c) electrons   d) helium nuclei

25. Equations describing nuclear reactions must balance

       a) charge   b) mass   c) atoms   d) both charge and mass

26. After a nuclear decay, a particle is 4 mass units lighter than it was previously. It most likely

       a) lost a neutron   b) gained a proton
       c) lost an alpha particle   d) gained a positron

27. In this equation, $_{13}^{24}Al \rightarrow$ _____ $+ _{12}^{24}Mg$, the missing product is

       a) an alpha particle   b) a beta particle   c) an electron   d) a positron

28. A material has a half-life of 2 minutes. If one begins with 160 grams and waits for 8 minutes, how many grams will remain?

       a) 40 g   b) 20 g   c) 10 g   d) 5 g

29. Which of these statements does *not* describe nuclear fusion?

       a) This reaction occurs at very high temperatures.
       b) This reaction uses uranium as a fuel.
       c) This reaction converts mass into energy.
       d) This reaction does not occur naturally on Earth.

30. Breeder reactors

       a) manufacture fuel   b) make use of nuclear fission
       c) use uranium as a starting material   d) All of the above.

31. If $_{82}^{210}Pb$ loses a beta particle, the resulting nuclide is

       a) $_{83}^{209}Bi$ b) $_{81}^{210}Ti$ c) $_{80}^{206}Hg$ d) $_{83}^{210}Bi$

32. In the equation $_{83}^{209}Bi + ? \rightarrow {}_{84}^{210}Po + _0^1n$, the missing bombarding particle would be

       a) $_1^2H$ b) $_0^1n$ c) $_2^4He$ d) $_{-1}^0e$

33. $_{94}^{241}Pu$ successively emits β, α, α, β, α, α. At that point, the nuclide has become

       a) $_{94}^{225}Pu$ b) $_{88}^{225}Pu$ c) $_{84}^{207}Po$ d) $_{84}^{219}Po$

34. Calculate the nuclear binding energy of $^{56}_{26}$Fe. Mass data: $^{56}_{26}$Fe = 55.9349 g/mol; n = 1.0087 g/mol; p = 1.0073 g/mol; e⁻ = 0.00055 g/mol; 1.0 g = 9.0 x 10^{13} J

    a) 4.8 x 10^{13} J/Mol   b) 56.4651 g/mol   c) 0.5302 g/mol   d) 4.9 x 10^{15} J/mol

35. The nuclide that has the longest half-life is

    a) $^{238}_{92}$U   b) $^{210}_{82}$Pb   c) $^{234}_{90}$Th   d) $^{222}_{88}$Ra

36. Which of the following is not a unit of radiation?

    a) Curie   b) Roentgen   c) Rod   d) Rem

37. What type of radiation is a very energetic form of photon?

    a) Alpha   b) Beta   c) Gamma   d) Positron

38. After emission of an alpha particle, the new daughter particle

    a) is two mass units lighter than before.
    b) is exactly the same mass as before emission.
    c) is actually two mass units heavier than before.
    d) is four mass units less than originally.

39. If a radioactive element emits one alpha and one beta particle, the new daughter particle

    a) has the same mass as before emission.
    b) has less mass than the original particle.
    c) has more mass than the original particle.
    d) has a mass which is 2 amu less than the original mass.

40. The half-life of a radioactive isotope of mercury is 31 days. What mass of a 0.100 g sample of this isotope will remain after 124 days have elapsed?

    a) 0.00625 g   b) 0.0125 g   c) 0.0250 g   d) 0.000 g

41. The half-life of ^{214}Bi is 19.7 minutes. Starting with 0.001 g of the bismuth isotope, how many grams will remain after 59.1 minutes?

    a) 1.25 x 10^{-4} g   b) 2.50 x 10^{-4} g   c) 3.33 x 10^{-4} g   d) 5.00 x 10^{-4} g

42. The half-life of radioactive ^{55}Cr is 1.8 hours. The delivery of a sample of this isotope from the reactor to a laboratory requires about 10.8 hours. What is the minimum amount of material that should be shipped in order that you receive 1.0 milligram of the chromium-55?

    a) 128 mg   b) 64 mg   c) 32 mg   d) 11 mg

43. Each of the following nuclear transformations would cause the mass number of a particular isotope to increase except

    a) beta emission   b) proton absorption   c) alpha absorption   d) neutron decay

44. Which of the following nuclear emissions has the highest penetrating power?

    a) alpha emission   b) beta emission   c) positron emission   d) gamma emission

45. Radioactivity has been used for all of the following except

    a) irradiation of food   b) medical diagnosis
    c) radiocarbon dating   d) creating new stable elements

46. In a nuclear reactor, the purpose of control rod is

    a) to convert the steam to electrical power
    b) to capture neutrons to slow down the rate of fission
    c) to cool the steam generated by the reactor
    d) to produce neutrons to speed up the rate of fission

47. Which is true about nuclear fusion?

    a) It requires very high temperature   b) Small nuclei are fused together
    c) It is the energy source of the sun   d) All are true.

48. Which is not true about nuclear fusion?

    a) Two or more neutrons are produced from the fission of each atom
    b) Most nuclides produce are stable nuclei
    c) The mass required to achieve a chain reaction is called the critical mass
    d) Vast amounts of energy are produced

49. The mass defect in the formation of an alpha particle from two protons and two neutrons is 0.0305 g/mol. The nuclear binding energy for this alpha particle is

    a) 2.7×10^{12} J/mol   b) 0.0305 g/mol   c) 9.0×10^{13} J/g   d) 2.9×10^{35} J/mol

50. Which is true about ionizing radiation?

    a) It dislocates bonding electrons and creates ions   b) It can damage DNA molecules
    c) Large acute dose or a chronic small dose are both harmful   d) All are true.

51. Which one of the following nuclear emissions has the largest mass number?
    a) alpha emission   b) beta emission   c) positron emission   d) gamma emission

Matching

From the list of terms given choose the one that correctly identifies each phrase.

List of terms: (a) Marie Curie; (b) Irene Joliot-Curie; (c) Hahn and Strassmann; (d) Antoine Henri Becquerel; (e) Paul Villard; (f) Edwin McMillan; (g) E. O. Lawrence; (h) Ernest Rutherford; (i) Wilhelm Roentgen.

_____ **1.** Discovered radioactivity in uranium salts.

_____ **2.** Discovered two new elements, polonium and radium, both of which are radioactive.

_____ **3.** Discovered the gamma ray.

_____ **4.** Performed the first man-made transmutation, converting nitrogen into oxygen.

_____ **5.** Prepared the first artificial radioisotope, $^{30}_{15}P$.

_____ **6.** Invented the cyclotron.

_____ **7.** Performed the first man-made nuclear fission by bombarding uranium with neutrons.

_____ **8.** Discovered the first transuranium element.

CHAPTER 19.   Introduction to Organic Chemistry

True — False (choose one)

1. The first organic compound synthesized from inorganic sources was urea.

2. Organic chemistry deals with the compounds of carbon.

3. The major sources of organic compounds are natural gas, petroleum and sea water.

4. The main type of bonding in organic compounds is ionic.

5. Carbon atoms can form as many as four covalent bonds with other atoms.

6. Carbon can form single, double, triple, and quadruple bonds between two carbon atoms.

7. The main reason for the existence of millions of organic compounds is the ability of carbon atoms to bond together into long chains and rings.

8. Hydrocarbons contain the elements carbon, hydrogen, and oxygen.

9. The difference between paraffins and saturated hydrocarbons is the presence of one or more double bonds in the paraffin molecules.

10. The common classes of organic compounds exist in homologous series.

11. Organic compounds having the molecular formula C_3H_8O will have similar properties.

12. Alkane molecules have little intermolecular attraction and therefore relatively low boiling points.

13. The electronegativity difference between carbon and hydrogen accounts for the marked polarity of alkane molecules.

14. Ethanol, CH_3CH_2OH, and dimethyl ether, CH_3—O—CH_3, are isomers.

15. Isomers are compounds that have the same structural formulas but different molecular formulas.

16. The alkanes are a homologous series of hydrocarbons.

17. The formula for isobutane is $(CH_3)_2CHCH_3$.

18. The name of the structure shown is 4-ethyl-2,2,4-trimethylheptane.

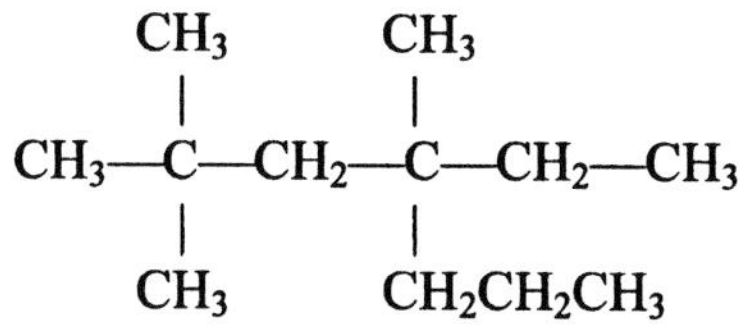

$$
\begin{array}{ccccc}
& CH_3 & & CH_3 & \\
& | & & | & \\
CH_3 & -C-CH_2- & & C-CH_2-CH_3 \\
& | & & | & \\
& CH_3 & & CH_2CH_2CH_3 &
\end{array}
$$

19. The name of the structure shown is 2,2,4-trimethyl-4-propylhexane.

$$\begin{array}{ccccc} & CH_3 & & CH_3 & \\ & | & & | & \\ CH_3 - & C - CH_2 - & C - CH_2 - CH_3 \\ & | & & | & \\ & CH_3 & & CH_2CH_2CH_3 \end{array}$$

20. A functional group establishes the identity of a class of compound and determines its chemical properties.

21. The reaction of an organic compound takes place at the functional groups.

22. There are two monochloro derivatives each for propane and butane.

23. Two names for $CH_3CH_2CHClCH_3$ are 2-chlorobutane and sec-butyl chloride.

24. Isopropyl chloride is a secondary alkyl halide.

25. When pentane is halogenated with chlorine, four monochloro products are possible.

26. The name for CH_3CCl_3 is 1,1,1-trichloroethane.

27. There are three isomeric pentanes.

28. $C_{12}H_{24}$ is a member of the alkane series.

29. Incomplete combustion of an alkane yields CO_2 and H_2O.

30. Alkenes and alkynes are both unsaturated hydrocarbons.

31. The simplest alkene and alkyne both contain the same number of carbon atoms.

32. $\begin{array}{c} CH_3 \\ | \\ CH_3CH_2CH=CH \end{array}$ is named 1-methyl-1-butene.

33. The open chain compound C_7H_{12} can have two double bonds or one triple bond in its structure.

34. The formula that represents octyne is C_8H_{16}.

35. The name for $CH_3C = CHCH_2CH_3$ is 2-ethyl-2-pentene.
$$\begin{array}{c} | \\ CH_2CH_3 \end{array}$$

36. When bromine is reacted with $CH_3CH_2CH = CH_2$ the product is $CH_3CH_2CHBrCH_2Br$.

37. When HCl is added to $CH_3CH = CHCH_3$ the product is $CH_3CHClCHClCH_3$.

38. The reaction $C_4H_{10} + Br_2 \rightarrow C_4H_9Br + HBr$ illustrates an addition type reaction.

39. Polymers are macromolecules.

40. The process of forming very large high-molar-mass molecules from smaller units is called polymerization.

41. Ethylene is the monomer of the polymer polyethylene.

42. Monomer molecules that unite to form polymers may be all the same or they may be different.

43. The monomer of polyvinyl chloride is $CH_2 = CHCl$.

44. Benzene compounds are frequently called aromatics.

45. There are six possible isomers in dichlorobenzene, $C_6H_4Cl_2$.

46. There are three possible isomers in tribromobenzene, $C_6H_3Br_3$.

47. There are three isomers of dibromobenzene, $C_6H_4Br_2$.

48. The name for is *m*-nitroaniline.

49. The name for is 1,5-dibromotoluene.

50. The name for is 1,2,3,4-tetrachlorobenzene.

51. The general formula for an alcohol is ROR.

52. $CH_3CH_2CHCH_2CH_2CH_3$ is a tertiary alcohol. Its name is 3-hexanol.
  |
  OH

53. The IUPAC name for $CH_3CHCH_2CHCHCH_3$ is 2,5-dimethyl-3-hexanol.

54. Ethylene glycol is widely used as an antifreeze.

55. When ingested in large quantities, ethyl alcohol acts as a poison and can cause depressed brain function, vomiting, and impaired perception.

56. A substance whose formula is $C_5H_{11}OH$ is a hexyl alcohol.

57. There are four isomeric butyl alcohols, C_4H_9OH.

58. Cetyl alcohol is a 16 carbon saturated alcohol. Its formula is $C_{16}H_{33}OH$.

59. Alcohols and ethers are isomeric compounds.

60. The general formula for ethers is R—O—R.

61. The number of possible isomers of C_3H_8O is four.

62. Ethers have higher boiling points than their isomeric alcohols.

63. Two names for CH_3CH_2—O—$CH_2CH_2CH_2CH_3$ are ethyl butyl ether and 1-ethoxybutane.

64. The formula for ethyl isopropyl ether is CH_3CH_2—O—$CH_2CH_2CH_3$.

65. The functional group for aldehydes and ketones is $\diagdown C = O.$

66. RCHO is a ketone and R_2CO is an aldehyde.

67. The name for $CH_3\overset{\overset{\displaystyle CH_3}{|}}{C}H\ CH_2\overset{\overset{\displaystyle H}{|}}{C}=O$ is 3-methylpentanal.

68. Two names for $CH_3CH_2\underset{\underset{\displaystyle O}{\|}}{C}CH_2CH_3$ are 3-pentanone and diethyl ketone.

69. Methyl ethyl ketone and butanal are isomers.

70. Formaldehyde is $H_2C = O$. Its largest use is for making polymers.

71. $CH_3CH_2CH_2CH_2 - \underset{\underset{\displaystyle O}{\|}}{C} - \bigcirc$

is called phenyl pentyl ketone.

72. The functional group of carboxylic acids is -COOH.

73. The simplest aromatic acid is benzoic acid.

74. The first three members of the carboxylic acid homologous series are ethanoic, propanoic, and butanoic acids.

75. The carboxylic acids are weak acids, and therefore cannot be neutralized by bases.

76. Another name for methanoic acid, HCOOH, is formal acid.

77. The formula for paimitic acid is $CH_3(CH_2)_{16}COOH$.

78. Esters have the general formula RCOOR′.

79. The lowest molar mass ester contains two carbon atoms.

80. Esters are alcohol derivatives of carboxylic acids.

81. Two names for HC—O—CH$_2$CH$_2$CH$_3$ are propyl methanoate and propyl acetate.
     $\parallel$
     O

82. Ethyl acetate can be formed by heating acetic acid and ethanol in the presence of sulfuric acid.

83. The name for CH$_3$CH$_2$C—OCH$_2$CH$_3$ is ethyl propanoate.
     $\parallel$
     O

84. The name for CH$_3$CH$_2$C–OCH$_2$CH$_3$ is ethyl acetate.
     $\parallel$
     O

85. Chlorination of CH$_3$CH$_2$CH$_2$CH$_3$ produces three monochloro substitution products.

86. The carbonyl group is —C—OH.
     O
     $\parallel$

87. The name of (CH$_3$)$_3$CBr is *tert*-butyl bromide or 2-bromo-2-methylpropane.

88. Dimethyl ketone and propanal have the same molar mass.

89. A compound of formula C$_6$H$_{12}$O cannot be a carboxylic acid.

90. Acetylene is the common name for ethyne.

91. Although ethyl alcohol is used in beverages, it is classified physiologically as a depressant and a poison.

92. Dichlorobenzene, C$_6$H$_4$Cl$_2$, has the same number of isomers as bromochlorobenzene, C$_6$H$_4$BrCl.

93. Another name for stearic acid is octadecanoic acid.

94. 2-butanone and butanal are isomers.

Multiple Choice (choose the best answer)

1. The number of covalent bonds a carbon atom forms is generally

    a) 2   b) 4   c) 6   d) 8

2. The most common geometric arrangement of bonds about a carbon atom is

    a) tetrahedral   b) linear   c) bent   d) planar

3. The most frequently encountered bond angle in organic compounds is approximately

a) 90°   b) 180°   c) 109°   d) 60°

4. Which formula represents an alkane?

a) C_nH_{2n} b) C_nH_{2n+2} c) C_nH_{2n-2} d) C_nH_n

5. Which formula represents an alkene?

a) C_nH_{2n} b) C_nH_{2n+2} c) C_nH_{2n-2} d) C_nH_n

6. Which formula represents an alkyne?

a) C_nH_{2n} b) C_nH_{2n+2} c) C_nH_{2n-2} d) C_nH_n

7. Which compound is not a homologue of the others?

a) C_6H_{14} b) C_5H_{10} c) CH_4 d) C_3H_8

8. Which one of the following is an alkane?

a) C_6H_6 b) C_6H_{14} c) C_6H_{12} d) C_6H_{10}

9. All of the following are true for alkanes except

a) Alkanes are non polar compounds
b) The boiling points of alkanes decrease with increasing chain length
c) Alkanes are soluble in non polar solvent
d) Alkanes are limited in reactivity

10. The difference in molar mass between two successive members of a homologous series is

a) 12   b) 16   c) 14   d) 15

11. Which formula represents a compound different from the other three?

a) $CH_3CH(CH_3)_2$ b) $CH_3CH_2CH_2CH_3$ c) $CH_3CH_2CH_2$ (with CH_3 branch below)   d) $CH_3(CH_2)_2CH_3$

12. Which name is not an IUPAC name?

a) ethane   b) ethene   c) ethyne   d) methyne

13. Which structure is the tert-butyl group?

a) CH_3CCH- (with CH_3 above)   b) CH_3-C- (with CH_3 above and CH_3 below)   c) CH_2-CHCH_3 (with CH_3 above and below)   d) CH_3CHCH_2 (with CH_3 above)

14. Which formula is a pentane?

    a) C_3H_8 b) C_4H_{10} c) C_5H_{10} d) C_5H_{12}

15.

$$CH_3$$
$$|$$
The name for $CH_3—C—CH_2CH_2CHCH_3$
$$\quad\quad\quad\quad | \quad\quad\quad\quad |$$
$$\quad\quad CH_2CH_3 \quad CH_2CH_3$$

    a) 2,5diethyl-2-methylhexane   b) 2-ethyl-2,5-dimethylheptane
    c) 3,3,6-trimethyloctane   d) No correct answer given.

16. Isomers are

    a) Compounds with the same molecular formula, but made with different isotopes.
    b) Compounds with the same molecular formula, but different molecular shapes.
    c) Compounds with the same empirical formula, but different molecular formula.
    d) Different compounds with the same empirical formula.

17. The number of isomers of butane is

    a) 1   b) 2   c) 3   d) 5

18. The number of isomers of pentane is

    a) 1   b) 2   c) 3   d) 5

19. How many monochloro substitution isomers are possible for hexane?

    a) 2   b) 3   c) 4   d) 6

20. How many dichloro derivatives of propane are possible?

    a) 3   b) 4   c) 5   d) 7

21. The reaction: $CH_3CH_3 + Cl_2 \rightarrow CH_3CH_2Cl + HCl$ is classified as

    a) substitution   b) addition   c) single displacement   d) oxidation

22. Which compound is chloroform?

    a) CH_3Cl b) CH_2Cl_2 c) $CHCl_3$ d) CCl_4

23. The name for $CH_3CH—CH=C—CH_3$ is
$$\quad\quad\quad\quad\quad\quad\quad | \quad\quad\quad |$$
$$\quad\quad\quad\quad\quad\quad CH_3 \quad\quad CH_3$$

    a) dimethyl-2-pentene   b) 2,4-dimethyl-2-pentene
    c) isoheptene   d) 2,4-dimethyl-3-heptene

24. An open chain hydrocarbon of formula C_7H_{10} can have in its formula

    a) one carbon-carbon double bond
    b) two carbon-carbon double bonds
    c) one carbon-carbon triple bond
    d) one carbon-carbon double bond and one carbon-carbon triple bond

25. Bromine reacts with unsaturated hydrocarbons by a process called

    a) polymerization   b) substitution   c) addition   d) ionization

26. In the reaction: $CH_3CH=CH_2 + HCl \rightarrow$, the organic product is

    a) $CH_3CHClCH_2Cl$ b) $CH_3CHClCH_3$ c) $CH_3CH_2CH_3$ d) $CH_2ClCHClCH_2Cl$

27. In the reaction: $CH_3CH+CHCH_3 + Br_2 \rightarrow$, the organic product is

    a) $CH_3CHBrCHBrCH_3$ c) $CH_3CH = CHCH_2Br$

    c) $CH_3CH_2CHBrCH_3$ d) $CH_2BrCH = CHCH_2Br$

                              CH_3
                               |
28. The name of $CH_3C{\equiv}C{-}CH_2CHCH_2CH_3$ is

    a) 5-methyl-2-heptyne   b) 3-methyl-5-heptyne
    c) isooctyne   d) methyl isopentyl acetylene

29. In the reaction: $CH_3CH{\equiv}CH + 2 Br_2 \rightarrow$, the organic product is

    a) $CH_3CBr=CHBr$ b) $CH_3CH=CHBr_2$ c) $CH_3CBr_2CHBr_2$ d) $CH_2BrCBr=CHBr$

30. Which is an incorrect formula?

    a) $CH_3CH_2CH_2Br$ b) $CH_3CHBr = CH_2$ c) $CH_2BrCH_2CH_2Br$ d) $CH_3CH = CHCH_2Br$

31. Which compound is the monomer of polyethylene?

    a) $CH_2 = CHCl$ b) $CH_2 = CH_2$ c) $CH_3CH = CH_2$ d) $CCl_2 = CCl_2$

32. How many ethylene units are in a polyethylene molecule that has a molar mass of 50,000?

    a) 1786   b) 2083   c) 3571   D) insufficent data to determine

33. When polymerized, the monomer $CF_2=CF_2$ polymerizes to

    a) polystyrene   b) orlon   c) teflon   d) polypropylene

34.   Simultaneous polymerization of $CH_2=CH_2$ and $CH_2=CHCl$ gives

      a) a polyethylene
      b) a polyvinylchloride
      c) a copolymer of ethylene and ethylenechloride
      d) a solution of polyethylene and polyvinylchloride

35.   The parent substance of the aromatic hydrocarbons is

      a) toluene   b) aniline   c) benzene   d) phenol

36.   Which two formulas cannot possibly represent the same substance?

      a) C_6H_6 and CH   b)  and

      b)  and C_6H_6 d)  and

37.   Which two formulas represent the same compound?

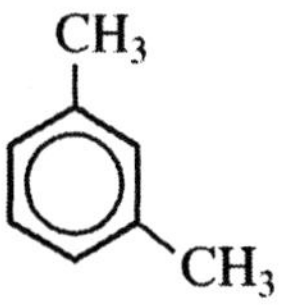

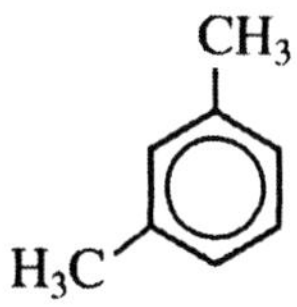

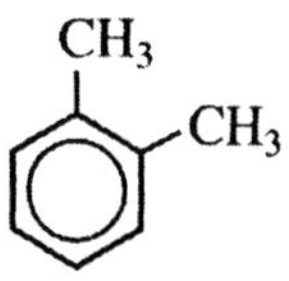

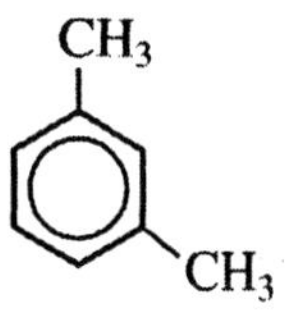

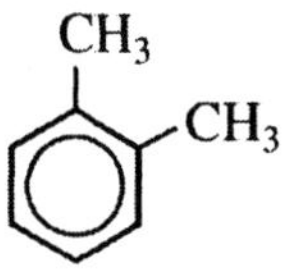

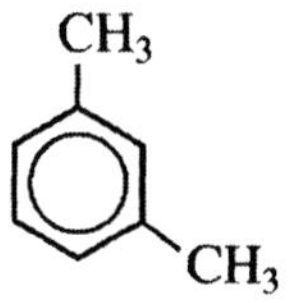

38.   In the compound

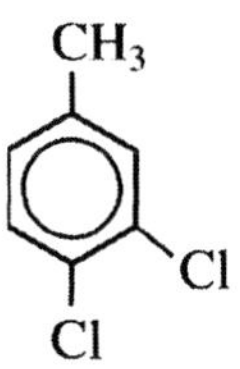

      a) both chlorine atoms are *para* to the CH_3 group and *ortho* to each other
      b) the methyl group is *ortho* to one chlorine and *para* to the other chlorine
      c) the chlorine atoms are *meta* to each other
      d) one chlorine atom is *para* to the methyl and *ortho* to the other chlorine

39. Which two names represent different compounds?

    a) toluene and methylbenzene
    b) 1,4-dibromobenzene and *para*-dibromobenzene
    c) phenol and hydroxybenzene
    d) 1,2-dichlobenzene and *meta*-dichlobenzene

40. In the structure, there is (are) with respect to X

    a) one *meta* position   b) one *para* position
    c) one *ortho* position   d) one *ortho* and one *para* position

41. Which compound is phenol?

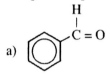

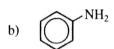

a)   b)

c)   d)

42. The name of   is

    a) 4-nitro-6-chlorotoluene   b) 2-chloro-4-nitrotoluene
    c) 1-methyl-2-nitrochlorobenzene   d) o,p-chloronitrotoluene

43. All of the following are functional groups except

    a) alcohol   b) Ether   c) Aromatic   d) Carboxilic acid

For questions 44 – 51 select answers from the following compounds:

    a. CH_3CH_2OH b. CH_3CHO c. CH_3COOH d. $CH_3CH(OH)CH_3$

44. Which compound is an aldehyde?

45. Which compound is a secondary alcohol?

46. Which compound will turn blue litmus red?

47. Which compound will react with methanol to form an ester?

48. From which compound could a ketone be easily made?

49. Which compound does not contain a hydrogen-oxygen bond?

50. Which compound contains the highest percent oxygen by mass?

51. Which compound contains the highest percent hydrogen by mass?

52. The number of isomers of butyl alcohol, C_4H_9OH, is

    a) 6   b) 5   c) 4   d) 2

53. Alcohols are isometric with

    a) ethers   b) aldehydes   c) ketones   d) carboxylic acids

54. The name for CH3(CH2)8CH2OH is

    a) 1-decanol   b) caprylic alcohol   c) decanal   d) 1-octanol

55. Which of the following is not a secondary alcohol?

    a) CH_3CHCH_3 b) $CH_3CH_2CHCH_3$ c) $CH_3CH_2CH_2OH$ d) CH_3CHCH_2OH
        | | |
        OH   OH   OH

56. Which ether is named 2-ethoxybutane?

    a) CH_3—O—$CHCH_3$ b) CH_3CH_2—O—$CHCH_2CH_3$
                   | |
                  OH   OH

    c) CH_3CH_2—O—$CHCH_3$ d) CH_3CH_2—O—CH_2CH_3
                   |
                  CH_3

57. The general formula for an aldehyde is

                                                      H
                                                    |
    a) R—O—R   b) RCOOH   c) R—C—R   d) R—C=O
                                        ‖
                                        O

58. The general formula for a ketone is

                                                      H
                                                    |
    a) R—O—R   b) RCOOH   c) R—C—R   d) R—C=O
                                        ‖
                                        O

59.   Another name for acetaldehyde is

   a) methanal   b) ethanal   c) propanal   d) butanal

60.   The compound $C_6H_{12}O$ cannot be a(n)

   a) alcohol   b) aldehyde   c) ketone   d) carboxylic acid

61.   The proper name for

$$CH_3CH_2CH_2CH\overset{\overset{\displaystyle CH_2CH_3}{|}}{\;}—CH_2CH\overset{\overset{\displaystyle H}{|}}{\underset{\underset{\displaystyle CH_3}{|}}{\;}}—C=O \text{ is}$$

   a) 2-methyl-4-propylhexanal   b) 4-ethyl-2-methylheptanal
   c) ethyl methyl caproaldehyde   d) 4-ethyl-2-methylheptanol

62.   The formula for propyl isopropyl ketone is

   a) $CH_3CH\underset{\underset{\displaystyle CH_3}{|}}{—}C\overset{\overset{\displaystyle }{||}}{\underset{\underset{\displaystyle O}{}}{—}}CH_2CH_3$ b) $CH_2CH_2\underset{\underset{\displaystyle CH_3}{|}}{—}C\overset{\overset{\displaystyle }{||}}{\underset{\underset{\displaystyle O}{}}{—}}CH_2CH_2CH_3$

   c) $CH_3CH\underset{\underset{\displaystyle CH_3}{|}}{—}C\overset{\overset{\displaystyle }{||}}{\underset{\underset{\displaystyle O}{}}{—}}CH_2CH_2CH_3$ d) $CH_3CH\underset{\underset{\displaystyle CH_3}{|}}{—}C\overset{\overset{\displaystyle }{||}}{\underset{\underset{\displaystyle O}{}}{—}}CH\underset{\underset{\displaystyle CH_3}{|}}{}CH_3$

63.   3,3-dimethylpentanoic acid contains how many carbon atoms?

   a) 5   b) 6   c) 7   d) 9

64.   Which of the following cannot be an aromatic carboxylic acid?

   a) C_6H_5COOH b) $C_6H_4(COOH)_2$ c) $C_6H_{13}COOH$ d) C_7H_7COOH

65.   Which of the following acids is named incorrectly?

   a) CH_3COOH, ethanoic acid   b) $CH_3(CH_2)_{16}COOH$, stearic acid

   c) $CH_3(CH_2)_8COOH$, decanoic acid   d) , methyl benzoic acid

66. The ester $CH_3\overset{O}{\overset{\|}{C}}-OCH_2\overset{CH_3}{\overset{|}{C}}HCH_3$ can be made from which alcohol and carboxylic acid?

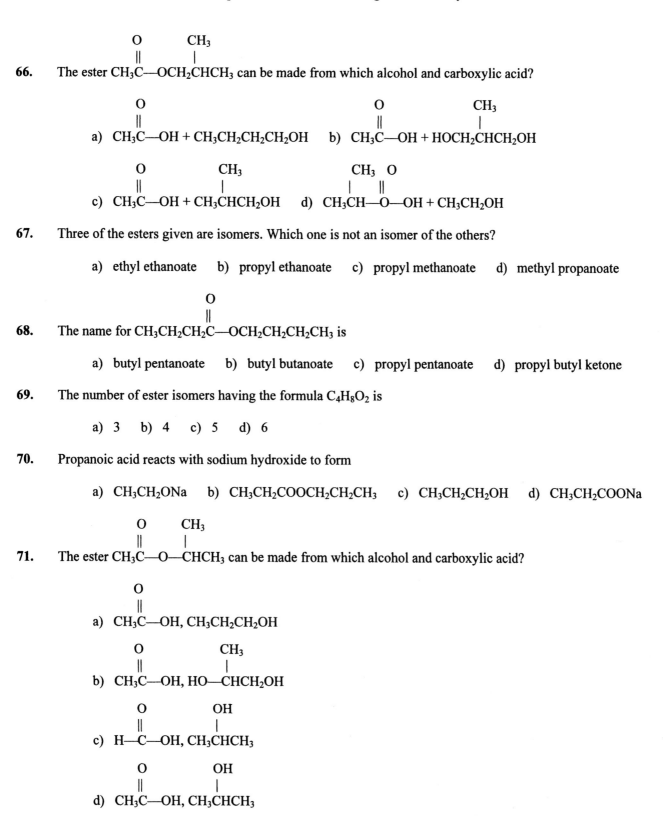

a) $CH_3\overset{O}{\overset{\|}{C}}-OH + CH_3CH_2CH_2CH_2OH$ b) $CH_3\overset{O}{\overset{\|}{C}}-OH + HOCH_2\overset{CH_3}{\overset{|}{C}}HCH_2OH$

c) $CH_3\overset{O}{\overset{\|}{C}}-OH + CH_3\overset{CH_3}{\overset{|}{C}}HCH_2OH$ d) $CH_3\overset{CH_3}{\overset{|}{C}}H-O-\overset{O}{\overset{\|}{}}OH + CH_3CH_2OH$

67. Three of the esters given are isomers. Which one is not an isomer of the others?

   a) ethyl ethanoate   b) propyl ethanoate   c) propyl methanoate   d) methyl propanoate

68. The name for $CH_3CH_2CH_2\overset{O}{\overset{\|}{C}}-OCH_2CH_2CH_2CH_3$ is

   a) butyl pentanoate   b) butyl butanoate   c) propyl pentanoate   d) propyl butyl ketone

69. The number of ester isomers having the formula $C_4H_8O_2$ is

   a) 3   b) 4   c) 5   d) 6

70. Propanoic acid reacts with sodium hydroxide to form

   a) CH_3CH_2ONa b) $CH_3CH_2COOCH_2CH_2CH_3$ c) $CH_3CH_2CH_2OH$ d) CH_3CH_2COONa

71. The ester $CH_3\overset{O}{\overset{\|}{C}}-O-\overset{CH_3}{\overset{|}{C}}HCH_3$ can be made from which alcohol and carboxylic acid?

   a) $CH_3\overset{O}{\overset{\|}{C}}-OH, CH_3CH_2CH_2OH$

   b) $CH_3\overset{O}{\overset{\|}{C}}-OH, HO-\overset{CH_3}{\overset{|}{C}}HCH_2OH$

   c) $H-\overset{O}{\overset{\|}{C}}-OH, CH_3\overset{OH}{\overset{|}{C}}HCH_3$

   d) $CH_3\overset{O}{\overset{\|}{C}}-OH, CH_3\overset{OH}{\overset{|}{C}}HCH_3$

72. With acid as a catalyst, ethanol and formic acid will react to form

a) $CH_3\overset{\overset{\displaystyle O}{\|}}{C}-NH_2$ b) $H-\overset{\overset{\displaystyle O}{\|}}{C}-O-CH_2CH_3$ c) $CH_3\overset{\overset{\displaystyle O}{\|}}{C}-O-CH_3$ d) $CH_3\overset{\overset{\displaystyle O}{\|}}{C}-O-\overset{\overset{\displaystyle O}{\|}}{C}CH_3$

73. Give the correct name for:

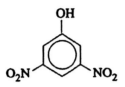

a) *m*-Dinitrophenol   b) 2,4-Dinitrophenol   c) 3,5-Dinitrophenol   d) 1,3-Dinitrophenol

74. The reaction of $CH_3CH=CHCH_3$ + HBr produces

a) $CH_3CH_2CH_2CH_3$ +Br   b) $CH_3CHBrCHBrCH_3$ + H_2
c) $CH_3CHBrCH_2CH_3$ d) $CH_3CH_2CH_2CH_2Br$

75. Polyvinyl chloride is a polymer of

a) $CH_2=CCl_2$ b) $CF_2=CF_2$ c) $CH_2=CHCl$ d) $C_6H_5CH=CHCl$

76. The correct name for

$$CH_3\overset{\overset{\displaystyle O}{\|}}{C}-O-\underset{\underset{\displaystyle CH_3}{|}}{\overset{\overset{\displaystyle CH_3}{|}}{C}}-CH_3 \text{ is}$$

a) acetyl-2-propanoate   b) propyl acetate   c) ethyl isopropylate   d) isopropyl ethanoate

77. Teflon is a polymer of

a) $CH_2=CCl_2$ b) $CF_2=CF_2$ c) $CH_2=CHCl$ d) $C_6H_5CH=CHCl$

CHAPTER 20.   Introduction to Biochemistry

True — False (choose one)

1.   The four major classes of biomolecules are carbohydrates, lipids, proteins, and sugars.

2.   The most abundant element in the human body is carbon.

3.   Biochemistry is concerned with the chemical reactions in living organisms.

4.   A monosaccharide is a carbohydrate which cannot be hydrolyzed to simpler carbohydrate units.

5.   Ribose, glucose, galactose, and sucrose are monosaccharides.

6.   The monosaccharide unit common to sucrose, maltose, and lactose is glucose.

7.   Lactose is commonly known as milk sugar.

8.   Sixteen different isomeric aldohexoses of formula $C_6H_{12}O_6$ are known, glucose being the most important one.

9.   Maltose is composed of two glucose units.

10.   Lactose is composed of a glucose unit and a ribose unit.

11.   Sucrose is a disaccharide composed of a glucose unit and a fructose unit.

12.   Cellulose and starch are polysaccharides containing many glucose units.

13.   The disaccharides have the formula $C_{12}H_{22}O_{11}$.

14.   Sucrose is the sweetest of the common sugars.

15.   Lipids have no common chemical structure.

16.   Fats and oils are esters of glycerol and fatty acids.

17.   Fats contain a higher percentage of unsaturated fatty acids than oils.

18.   Fats and oils are known as triacylglycerols or triglycerides.

19.   Linoleic acid is an unsaturated fatty acid which contains two carbon-carbon double bonds.

20.   When metabolized, fats generally produce about three times the energy per gram as do carbohydrates.

21.   Oleic, linoleic, and arachidonic acids are essential fatty acids.

22.   When a fat or an oil is saponified by reaction with sodium hydroxide, the products are glycerol and soap.

23. The formula for a typical soap is C_3H_7COONa.

24. The formula for a typical soap is $C_{15}H_{31}COONA$.

25. Cholesterol is a steroid and is classified as a lipid.

26. Major food sources of cholesterol are meat, liver, and egg yolk.

27. Amino acids are the building blocks of lipids.

28. All amino acids contain carbon, hydrogen, oxygen, and nitrogen.

29. Proteins are high molar mass polymers of amino acids.

30. All naturally occurring amino acids are alpha-amino acids.

31. $$H_2N-\overset{\overset{\textstyle O}{\|}}{C}-OH$$ is an alpha amino acid.

32. Of the approximately 200 known amino acids, 25 of them are found in almost all proteins.

33. The bond connecting amino acids in polypeptides and proteins is called a peptide bond or linkage.

34. An octapeptide has seven peptide bonds.

35. A peptide chain is numbered starting with the C-terminal amino acid residue.

36. The main function of proteins is to supply heat and energy to the body.

37. The ultimate end-product of protein catabolism is carbon dioxide, water, and urea.

38. There are eight essential amino acids which the human body is not able to synthesize.

39. Essential amino acids are those which the human body needs but cannot synthesize.

40. The synthesis of proteins is controlled by nucleic acids.

41. DNA and RNA are nucleic acids.

42. Nucleic acids are polymeric substances made up of thousands of units called nucleotides.

43. A nucleotide contains one unit each of phosphoric acid, a pentose sugar, and a nitrogen containing base.

44. The pentose sugar in DNA is ribose.

45. The pentose sugar in RNA is ribose.

46. The structure of DNA as proposed by Watson and Crick is that of two polymeric strands of nucleotides in the form of a double helix.

47. The genetic code of life is contained in the nucleic acid RNA.

48. The double helix of DNA is held together by hydrogen bonds between complementary base units.

49. Two major differences in RNA and DNA are that RNA contains the pentose sugar ribose and uracil while DNA contains deoxyribose and thymine.

50. The main function of RNA is to direct the synthesis of proteins.

51. Genes are segments of DNA that contain the basic units of heredity and are located in the chromosomes.

52. The process whereby DNA replicates itself during cell division is called meiosis.

53. Saturated describes a fatty acid with carbon-carbon double bonds.

54. The body does not store free amino acids.

55. Glucose, fructose, and sucrose are all simple sugars.

56. The body stores glucose as glycogen until it is needed.

57. The ultimate use of carbohydrates in the body is oxidation to carbon dioxide and the utilization of the energy released.

58. Polysaccharides are macromolecules made up of many monosaccharide units linked together.

59. Maltose is a disaccharide composed of two glucose units.

60. Triacylglycerols are also called triglycerides.

61. Fats and oils are esters of glycerol and the higher molar-mass fatty acids.

62. Oils have a higher percentage of unsaturated fatty acids than fats.

63. Fats are burned in the body or stored as fats for future use.

64. Paimitic acid contains one carbon-carbon double bond.

65. When a fat is saponified, the products are glycerol and soap.

66. Saturated fats are composed entirely of single bonds between carbons.

67. Unsaturated fats are composed entirely of double bonds.

68. Amino acids are organic compounds containing at least two functional groups: an amino group and a carboxyl group.

69. $H_2N-\overset{\displaystyle O}{\overset{\displaystyle \|}{C}}-OH$ is the formula for urea.

70. The double helix of DNA is held together by hydrogen bonds between the complementary base units.

71. For any individual of any species, the sequence of base combinations and the length of the nucelotide chains on DNA molecules contain the coded messages that determine the characteristics of the individual.

72. In the polypeptide Ala-Glu-Asn-Arg-Gly-Gly, the *N*-terminal amino acid is alanine.

73. Ala-Glu-Asn-Arg-Gly-Gly has five peptide bonds and is, therefore, a pentapeptide.

74. Peptide bonds connect one amino acid to another in a protein.

75. Polypeptides are numbered starting with the C-Terminal amino acid.

76. Enzymes are proteins that block specific biochemical reactions.

Multiple Choice (choose the best answer)

1. Which carbohydrate classification does *not* apply to glucose?

a) aldose   b) ketose   c) monosaccharide   d) aldohexose

2. Which saccharide does *not* belong in the same classification as the other three?

a) maltose   b) glucose   c) sucrose   d) lactose

3. Which sugar is *not* a hexose?

a) ribose   b) glucose   c) fructose   d) galactose

4. Cellulose is an example of a

a) protein   b) carbohydrate   c) lipid   d) nucleic acid

5. Which is *not* true about fructose?

a) it is a component of sucrose   b) it is an aldohexose
c) it is the sweetest sugar   d) it is a monosaccharide

6. Which is *not* true about glucose?

a) it is also known as dextrose   b) it is a component of starch and cellulose
c) it is the monomer of glycogen   d) it is a pentose

7. The sweetest of the common sugars is

a) fructose   b) glucose   c) sucrose   d) lactose

8. The structure shown is

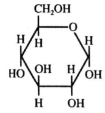

a) glucose   b) galactose
c) ribose   d) maltose

9. The products formed when maltose is hydrolyzed are

a) glucose and fructose   b) galactose and glucose
c) glucose and glucose   d) galactose and fructose

10. Which compound is a pentose?

a) ribose   b) glucose   c) galactose   d) fructose

11. Which is *not* a polysaccharide?

a) cellulose   b) starch   c) insulin   d) lactose

12. For a compound to be classified as a lipid, it must

a) contain a glycerol unit   b) be soluble in fat solvents
c) contain C, H, O, and N   d) be soluble in water

13. The formula $CH_3(CH_2)_{14}COOH$ represents

a) palmitic acid   b) stearic acid   c) oleic acid   d) lauric acid

14. Cholesterol is a

a) fat   b) glycolipid   c) phospholipid   d) steroid

15. Which acid is *not* unsaturated?

a) stearic   b) oleic   c) linoleic   d) linolenic

16. The number of ester groups per molecule of fat is

a) 1   b) 2   c) 3   d) 4

17. Which of the following is *not* a vegetable oil?

a) linseed   b) lard   c) olive   d) corn

18. Which is *not* formed in the saponification of a fat?

    a) glycerol   b) amino acids
    c) soap   d) a metal salt of a long-chain fatty acid

19. When a fat is saponified with caustic soda (NaOH) the products are

    a) sodium chloride and the fatty acids   b) sodium glycerate and the fatty acids
    c) glycerol and the sodium salts of the fatty acids   d) glycerol and sodium chloride

20. Which is *not* an essential fatty acid?

    a) oleic acid   b) linoleic acid   c) linolenic acid   d) arachidonic acid

21. Which is *not* a good source of protein?

    a) fish   b) nuts   c) eggs   d) wheat

22. An alpha amino acid always contains

    a) a carboxyl group at each end of the molecule
    b) two amino groups
    c) an amino group on the carbon adjacent to the carboxyl group
    d) alternating amino and carboxyl groups

23. A compound containing ten amino acid residues linked together is called a

    a) protein   b) deca-amino acid   c) polypeptide   d) nucleotide

24. Which statement is *not* true about the peptide Gly-Ala-Tyr-Phe?

    a) The molecule contains three peptide linkages.
    b) The N-terminal residue is glycine (Gly).
    c) Twelve different tetrapeptides can be written using each of these four amino acids only once.
    d) The name of the peptide is glycylalanyltyrosylphenylalanine.

25. Which of the following is *not* a correct statement about DNA and RNA?

    a) DNA contains deoxyribose, whereas RNA contains ribose.
    b) Both DNA and RNA are polymers made up of nucleotides.
    c) DNA contains the genetic code of life and RNA directs the synthesis of proteins.
    d) DNA exists as a single helix, whereas RNA exists as a double helix.

26. The double helical structure of DNA is held together by

    a) ionic bonds   b) hydrogen bonds   c) peptide bonds   d) phosphate ester bonds

27. Which of the following is *not* a fundamental component of DNA?

    a) deoxyribose   b) glycine   c) cytosine   d) phosphoric acid

28. The main function of RNA is to

    a) direct protein production   b) direct the hydrolysis of starches
    c) direct the synthesis of lipids   d) All of the above.

29. Which base is not found in DNA?

    a) adenine   b) cytosine   c) thymine   d) uracil

30. In a DNA double helix, which pair of bases are complementary, that is, held together by hydrogen bonding?

    a) cytosine and guanine   b) adenine and cytosine
    c) guanine and thymine   d) guanine and adenine

31. Which of the following scientists did not receive the Nobel Prize for the structure of DNA?

    a) Watson   b) Crick   c) Sanger   d) Wilkins

32. Proteins are composed of combinations of

    a) carbohydrates   b) fatty acids   c) lipids   d) amino acids

33. Lipid molecules consist of long chains of carbon atoms surrounded by hydrogen atoms. This basic structure explains why lipids have

    a) low solubility in water   b) soft texture
    c) high melting point   d) high chemical reactivity

34. The structure below shows a(n)

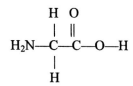

    a) alcohol   b) amine   c) amino acid   d) saturated fat

35. A protein which catalyzes a biochemical reaction is called

    a) a substrate   b) unsaturated   c) activated   d) an enzyme

36. Which is *not* true about starch?

    a) It is a polysaccharide.   b) It is hydrolyzed to maltose.
    c) It is composed of glucose units.   d) It is not digestible by humans.

37.   Lactose is

a) a monosaccharide   b) a disaccharide composed of galactose and glucose
c) a disaccharide composed of two glucose units   d) a decomposition product of starch

38.   Which is *not* formed in the saponification of a fat?

a) glycerol   b) amino acids   c) soap   d) a metal salt of a long-chain fatty acid

39.   Which is *not* an essential fatty acid?

a) oleic acid   b) linoleic acid   c) linolenic acid   d) arachidonic acid

40.   An alpha amino acid always contains

a)   An amino group on the carbon atom adjacent to the carboxyl group
b)   A carboxyl group at each end of the molecule
c)   Two amino groups
d)   Alternating amino and carboxyl groups

41.   A compound containing ten amino acid molecules linked together is called a

a) protein   b) polypeptide   c) deca-amino acid   d) nucleotide

42.   Which of the following statement is true?

a) proteins are one class of animal food
b) chromosomes are composed of proteins
c) proteins are polymers of amino acids
d) All the above are true.

43.   A maltose is

a) A disaccharide made from one unit of glucose and one unit of fructose
b) A disaccharide made from two units of glucose
c) A disaccharide made from one unit of galactose and one unit of glucose
d) None of the above.

44.   Which of the following statement is true regarding fat?

a) Fats are major constituents of adipose tissue
b) Fats insulate the body against loss of heat
c) Fats are solid at room temperature
d) All of the above.

45. All of the following statements are true regarding DNA except
   a) DNA contains the genetic code of life
   b) DNA contains thymine and deoxyribose
   c) Replication of DNA begins with the unwinding of the double helix at the hydrogen bonds between the base
   d) Ionizing radiation can irreversibly damage DNA

46. Which of the following is true regarding RNA?

   a) RNA exists in the form of a single-stranded helix
   b) RNA contains the pentose sugar ribose
   c) RNA contains the base uracil
   d) All of the above.

47. The major end-product(s) of protein nitrogen metabolism in humans is (are)

   a) amino acids   b) ammonium salts   c) urea   d) dipeptides

48. Which of the following bases is found in RNA, but not in DNA?

   a) thymine   b) adenine   c) guanine   d) uracil

49. In a DNA double helix, hydrogen bonding occurs between

   a) adenine and thymine   b) thymine and guanine
   c) adenine and uracil   d) cytosine and thymine

50. Which of the following scientists did *not* receive the Nobel Prize for the structure of DNA?

   a) Crick   b) Watson   c) Miescher   d) Wilkins

51. The substance acted on by an enzyme is called a(n)

   a) catalyst   b) apoenzyme   c) coenzyme   d) substrate

Answers to Chapter Tests (pages 1–75)

Chapter 1 Test

True - False: 1. T 2. T 3. F 4. F 5. T

Multiple Choice: 1. b 2. d 3. a 4. c 5. d 6. d

Reasoning and Expression:

1. (a) Buehler collected facts and data through experimentation, (b) he proposed a hypothesis regarding the two phases of nitinol, (c) he and his group continued to test their hypothesis until they understood the mechanism of nitinol well.

2. A hypothesis becomes a theory after it has been repeatedly tested and found to be consistent with evidence and facts; a law is a theory, in a sense, to which no exception has been found under given circumstances.

3. Many answers are possible—among them are: does it dissolve in water, is it a solid or a liquid or a gas, what is its melting point or boiling point, what is its density, and what elements comprise it?

4. It is important to know if the danger of a chemical will ever be encountered (probability) and, if so, what the likely consequences (severity) may be.

5. (a) Collect the facts and/or data, (b) formulate a hypothesis, (c) test the hypothesis, and (d) modify the hypothesis as needed to fit the observed and pertinent data.

6. Many answers are possible. One could organize by those with flowers and those without, by the shape of the leaf, by the height of the plant, by those that grow best in the shade and those in the sun.

Chapter 2 Test

True - False: 1. T 2. F 3. T 4. T 5. T 6. T 7. F 8. F 9. F 10. T 11. T

Multiple Choice: 1. d 2. c 3. a 4. a 5. b 6. a 7. d 8. c 9. a 10. d 11. c

Reasoning and Expression:

1. 25,800 g
2. 350K
3. 0.75 g/cm^3, it will float since its density is less than that of water
4. (a) mm (b) mL (c) km (d) joule (e) m^2
5. Heat measures energy contained in all the molecules in a sample collectively; temperature measures the intensity of that heat as demonstrated by the molecules directly around the thermometer. Temperature does not depend upon the size of the sample.

6. (a)

(b) About 105 buttons. (c) (20.8 + 82.0 + 35.8 g)/200 button ~ 0.69 g/button

Chapter 3 Test

True - False: 1. T 2. T 3. T 4. T 5. F 6. T 7. F 8. F 9. F 10. F 11. T

Multiple Choice: 1. a 2. c 3. b 4. b 5. c 6. a 7. b 8. c 9. b 10. c 11. a 12. b

Reasoning and Expression:

1. 6 in $Ca(NO_3)_2$, 16 in 8 $Ca(OH)_2$, 36 in 12 $CaCO_3$, 48 in 6 $Ca_3(PO_4)_2$
2. (a) an ion has an electrical charge, (b) an anion has a negative charge, a cation has a positive charge, (c) a solid has a fixed volume and shape, a gas has neither, (d) a molecule is a cluster of two or more atoms
3. Many answers are possible—among them are: try to dissolve the powders in water, heat them to find melting points, detect any aroma, observe texture, burn them in air, and so on.
4. (a) 1 (b) 4 (c) 3
5. (a) $\frac{4}{5}$ (b) $\frac{52}{77}$ (c) $\frac{12}{24} = \frac{1}{2}$ (d) $\frac{4}{8} = \frac{1}{2}$
6. (a) He, Ne, Ar, Kr, Xe, Rn (b) I_2 (c) Hg (d) Au, Ag, Pt (e) Na

Chapter 4 Test

True - False: 1. F 2. T 3. F 4. T 5. T 6. F 7. T 8. T 9. T 10. F 11. T

Multiple Choice: 1. b 2. c 3. d 4. b 5. b 6. c 7. b 8. b 9. c 10. c

Reasoning and Expression:

1. Determine the difference in mass before and after the copper strip reacted (36 g - 12 g = 24 g), compare the mass difference (the amount of iodine which reacted) to the total mass of the compound (24g/36 g = 0.67), change that fraction to a percentage so that it becomes 67% I.
2. 57.9°C
3. (a) $Ca + 2H_2O \rightarrow Ca(OH)_2 + H_2$ (b) $Fe + S \rightarrow FeS$ (c) $C + O_2 \rightarrow CO_2$
4. Potential energy increases as the club is swung back, potential energy becomes kinetic energy as the club is brought down, the kinetic energy of the club transfers to kinetic energy of the ball when the two strike, and the potential energy of the club increases once again as the club is brought back on follow-through. Potential energy is maximum at the point of the original back swing. Kinetic energy is maximum right before the club hits the ball.
5. A substance with a low specific heat needs relatively little energy to change temperature.

270

6.

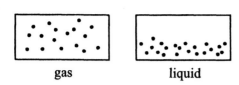

gas   liquid

7. The one that lost mass could have decomposed to emit a gas which was lost to the air, the one that gained mass could have chemically reacted with oxygen or some other component of the air; the one that didn't change perhaps just warmed or cooled. All still obeyed the Law of Conservation of Mass, if one includes the entire system: the dish, the air around the dish, and the chemicals that are contained therein.

Chapter 5 Test

True - False: 1. F   2. T   3. F   4. F   5. F   6. F   7. T   8. T   9. F   10. F   11. T

Multiple Choice: 1. c   2. c   3. a   4. d   5. a   6. a   7. c   8. c   9. b   10. c   11. b

Reasoning and Expression:

1. Dalton's model did not include the existence of positive charges, Thomson's model had positive charges scattered throughout the atom, and Rutherford's model placed the positive charge in the nucleus.
2. In Rutherford's alpha scattering experiments (a) alpha particles, which are positively-charged, are deflected (b) the alpha particles, which are relatively massive compared to an individual proton, were deflected at great angles (c) most of the alpha particles passed through the atom completely undeflected
3. 81.4 mnu
4. Most elements have one isotope which is vastly more common than any others.
5. They have the same mass number, though they represent different elements.
6. The few alpha particles that came near the nucleus (and in Rutherford's experiment veered sharply away) would have been attracted to and probably trapped inside the nucleus.

Chapter 6 Test

True - False: 1. T   2. T   3. T   4. F   5. F   6. F   7. T   8. F   9. F   10. T   11. T
                              12. F   13. F   14. F   15. F   16. T   17. F   18. T   19. T   20. F   21. F   22. F

Names and Formulas: 1. sulfuric acid; 2. nitric acid; 3. acetic acid; 4. calcium hydroxide; 5. iron (III) sulfate; 6. magnesium carbonate; 7. tin (II) sulfide; 8. HCl; 9. H_3PO_4; 10. KOH; 11. Na_2SO_4; 12. $AlCl_3$; 13. Cu_2O; 14. NH_4Br

Reasoning and Expression:

1. (a) $BaCl_2 + H_2SO_4 \rightarrow BaSO_4 + HCl$ (b) $Na_2CO_3 + Pb(NO_3)_2 \rightarrow NaNO_3 + PbCO_3$
(c) $Cu(NO_3)_2 + NaOH \rightarrow Cu(OH)_2 + NaNO_3$
2. (a) Q_2Z_5 (b) A_4B (c) $(ABC)_3(XYZ)_2$
3. (a) Toward the left of the periodic table. (b) Toward the right of the periodic table.
4. They form ions with a charge of +2. Column or group IIA. They form +1 ions.
5. Many answers are possible—among them are copper, iron, mercury, lead, tin, and chromium.
6. +2, +4, +1, +4

271

Chapter 7 Test

True - False: 1. T   2. T   3. F   4. T   5. T   6. T   7. T   8. T   9. T   10. F   11. T

Multiple Choice: 1. d   2. b   3. b   4. d   5. b   6. c   7. d   8. d   9. b   10. c   11. d

Reasoning and Expression:

1. Al weighs least. BF_3 has greatest number of atoms.
2. $N_2H_8CrO_4$
3. Density measures mass per volume. Atomic mass measures mass per number (mole). They both involve mass.
4. 4 carbon atoms
5. Yes, empirical. Subscripts cannot be reduced any further.
6. (a) $\left(\frac{8.0 \text{ g X}}{4.0 \times 10^{23} \text{ mol}}\right)\left(\frac{6.022 \times 10^{23} \text{ molecules}}{1 \text{ mol}}\right) = 120$ g/mol   (b) $(0.00842 \text{ kgY})\left(\frac{1000 \text{g}}{1 \text{ kg}}\right)\left(\frac{1}{0.203 \text{ mol}}\right) = 41.5$ g/mol

(c) 27.25 ml $\times \frac{5.82 \text{ g}}{1 \text{ mL}} \times \frac{1}{9.4 \times 10^{23} \text{ molecules}} \times \frac{6.02 \times 10^{23} \text{ molecules}}{1 \text{ mol}} = 102$ g/mol

Chapter 8 Test

True - False: 1. F   2. T   3. F   4. T   5. T   6. F   7. T   8. T   9. T   10. T   11. F

Multiple Choice: 1. d   2. a   3. d   4. c   5. b   6. c   7. d   8. d   9. b   10. b   11. d

Reasoning and Expression:

1. 16 atoms Al, 24 atoms S, 96 atoms O, 136 atoms altogether
2. (a) decomposition (b) energy is added and the molecule X_3Y breaks apart (c) the total number of atoms, the total mass of reactants and products, and the identity of each element.
3. (a) $4 Ag + 2 H_2S + O_2 \rightarrow 2 Ag_2S + 2 H_2O$
(b) $8 Al + 3 Fe_3O_4 \rightarrow 4 Al_2O_3 + 9 Fe$
4. $C_6H_6 + 3 HNO_3 \rightarrow 3 H_2O + C_6H_3N_3O_6$
5. A subscript is part of the empirical or molecular formula of the compound. A coefficient simply tells us how many of an atom or molecule are needed to balance an equation.
6. (a) $CaCO_3 \rightarrow CaO + CO_2$ (b) Step #3   (c) Step #1
$2 NH_4Cl + CaO \rightarrow 2NH_3 + H_2O + CaCl_2$
$CO_2 + NH_3 + H_2O \rightarrow NH_4HCO_3$
$NH_4HCO_3 + NaCl \rightarrow NaHCO_3 + NH_4Cl$

Chapter 9 Test

True - False: 1. T   2. T   3. T   4. F   5. T   6. T   7. T   8. F   9. T   10. T   11. F

Multiple Choice: 1. d   2. c   3. b   4. d   5. c   6. c   7. b   8. a   9. d   10. b   11. a

Reasoning and Expression:

1. $2 NO_2 \rightarrow N_2O_4$; 1000
2. $6 NO + 4 NH_3 \rightarrow 5 N_2 + 6 H_2O$; (a) $\frac{6}{6} = \frac{1}{1}$ (b) $\frac{5}{6}$ (c) $\frac{4}{6} = \frac{2}{3}$
3. 64 grams SO_2

4. The copper wire must have reacted with something in the air to form a compound on the wire.

5. (a) $C_6H_{12}O_6 \rightarrow 2\ C_2H_5OH + 2\ CO_2$ (b) 256 g alcohol (c) 1.30 liters

6. (a) $CaCO_3 \rightarrow CaO + CO_2$

(b) $(250.0\ CaO)\left(\frac{1\ mol\ CaO}{56.08\ g\ CaO}\right)\left(\frac{1\ mol\ CaCO_3}{1\ mol\ CaO}\right)\left(\frac{100.1\ g\ CaCO_3}{mol}\right)\left(\frac{100\%}{95\%}\right) = 470\ g\ CaCO_3$

7. (a) $Co + FeCl_2 \rightarrow Fe + CoCl_2$

(b) $(19.0\ g\ Co)\left(\frac{1\ mol\ Co}{58.93\ g\ Co}\right)\left(\frac{1\ mol\ FeCl_2}{1\ mol\ Co}\right)\left(\frac{126.8\ g\ FeCl_2}{1\ mol\ FeCl_2}\right) = 40.9\ g\ Fe$ needed (not enough).

$FeCl_2$ is the limiting reagent.

$(19.0\ g\ FeCl_2)\left(\frac{1\ mol\ FeCl_2}{126.8\ g\ FeCl_2}\right)\left(\frac{1\ mol\ Co}{1\ mol\ FeCl_2}\right)\left(\frac{58.93\ g\ Co}{1\ mol\ Co}\right) = 8.83\ g\ Co$ needed

$19.0\ g - 8.83\ g = 10.2\ g\ Co$ left over.

(c) $(19.0\ g\ FeCl_2)\left(\frac{1\ mol\ FeCl_2}{126.8\ g\ FeCl_2}\right)\left(\frac{1\ mol\ Fe}{1\ mol\ FeCl_2}\right)\left(\frac{55.85\ g\ Fe}{1\ mol\ Fe}\right) = 8.37\ g\ Fe$ formed

$(19.0\ g\ FeCl_2)\left(\frac{1\ mol\ FeCl_2}{126.8\ g\ FeCl_2}\right)\left(\frac{1\ mol\ CoCl_2}{1\ mol\ FeCl_2}\right)\left(\frac{129.8\ g\ CoCl_2}{1\ mol\ CoCl_2}\right) = 19.4\ g\ CoCl_2$ formed

Chapter 10 Test

True - False: 1. T   2. T   3. T   4. T   5. T   6. T   7. T   8. T   9. T   10. T   11. T

Multiple Choice: 1. c   2. a   3. b   4. d   5. d   6. d   7. b   8. c   9. a   10. d   11. a   12. a

Reasoning and Expression:

1.

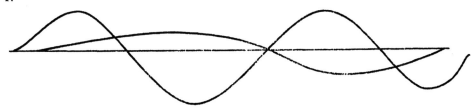

2. (a) H, He, Li, Be (b) Ti, V, Cr (c) N, P, As, Sb, Bi

3. (a) 0 (b) $\frac{6}{18} = \frac{1}{3}$ (c) $\frac{24}{55}$

4. In general, from top to bottom in a family, as atomic number increases, so does atomic radius. However, from left to right, as atomic number increases, there is a general and small decrease in size. From 18 to 19 electrons, a new energy level has been added; hence, a large increase in radius.

5. (a) 1 (b) 2 (c) 1 (d) 0

6. There are a number of ways to show evidence: infrared light gives the sensation of warmth, ultraviolet light causes tanning or sun burning.

Chapter 11 Test

True - False: 1. F 2. F 3. T 4. F 5. T 6. T 7. T 8. T 9. T 10. T 11. T

Multiple Choice: 1. a 2. b 3. b 4. c 5. c 6. d 7. c 8. b 9. c 10. a 11. d 12. a 13. c

Reasoning and Expression:

1. e<b<c<a<d

2. BH_3 is a trigonal planar   H:B:H / H BH_4^- is tetrahedral   H / H:B:H / H

3. (a) $\left[\begin{array}{c} :\ddot{O}:C::\ddot{O}: \\ :\ddot{O}: \end{array} \right]^{2-}$ (b) $\left[:\ddot{N}:N:::N: \right]^-$ (c) $:\ddot{C}l:C::\ddot{O}: \\ :\ddot{C}l:$

4. They all have one unpaired electron; they will be unstable.

5. (a) $\left[:\ddot{C}l: \right]^-$ (b) $\left[:\ddot{C}l: \right]^-$ (c) $:\ddot{C}l:S:\ddot{C}l:$

(b) Both involve two bonds; Mg-Cl is ionic and S-Cl is polar covalent.
(c) $MgCl_2$ has a crystalline structure; SCl_2 has an angular or bent structure.
(d) $MgCl_2$ is nonpolar, whereas SCl_2 is a dipole.
6. The molecule could be pyramidal or even T-shaped. The shape will be determined not only by the number of atoms surrounding the central atom but also by the number of nonbonding electron pairs that are present on the central atom.

Chapter 12 Test

True - False: 1. T 2. T 3. T 4. T 5. F 6. T 7. F 8. T 9. F 10. F 11. F

Multiple Choice: 1. d 2. b 3. a 4. c 5. c 6. b 7. b 8. c 9. c 10. a 11. a

Reasoning and Expression:

1. 5.3 g/L
2.

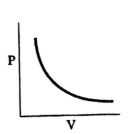

3. (a) heat causes molecules of carbon dioxide to expand (b) liquid water evaporates from your skin, absorbing energy in the process of changing from a liquid to a vapor.
4. 1.25 mL
5. 44 g/mol

6. d. The reason this is the answer is that, even though temperature and pressure change, gas molecules (as long as they remain gaseous) will still be equidistant from one another, bouncing rapidly and randomly.

Chapter 13 Test

True - False: 1. F   2. F   3. T   4. F   5. T   6. T   7. T   8. F   9. T   10. F   11. F

Multiple Choice: 1. b   2. c   3. d   4. d   5. b   6. a   7. a   8. c   9. a   10. b   11. a

Reasoning and Expression:

1. (a)

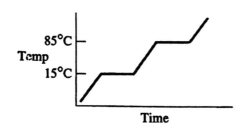

(b) 5600 kJ
(c) At 15°C the solid would melt, and at 85°C, you would see the liquid change to vapor. Beyond that temperature, you would see vapor continue to warm to 100°C.
2. (a) Melting point is the temperature at which something changes from solid to liquid; boiling point is where it changes from liquid to gas.
(b) Heat of fusion is the energy required to change 1 gram of a substance from a solid to a liquid at its melting point. Freezing point is the temperature at which the solid and liquid states of a substance are in equilibrium.
(c) Boiling changes a liquid to a vapor; subliming changes a solid directly to a vapor.
3. 330 g of ice
4. The plateau area on the curve would be extended as more substance was used.
5. (a) $SO_2 + H_2O \rightarrow H_2SO_3$ (b) $Li_2O + H_2O \rightarrow 2LiOH$ (c) $2Na + 2H_2O \rightarrow 2NaOH + H_2$
6. (a)

$$H-\overset{\overset{\displaystyle H}{|}}{\underset{\underset{\displaystyle H}{|}}{C}}-C\overset{O--------H-O}{\underset{O-H---------O}{\diagup}}C-\overset{\overset{\displaystyle H}{|}}{\underset{\underset{\displaystyle H}{|}}{C}}-H$$

(b) The hydrogen bonds cause a stronger connection to develop between molecules. Therefore, more energy is required to break them apart, and the subsequent boiling point is higher.

Chapter 14 Test

True - False: 1. F   2. T   3. T   4. T   5. T   6. T   7. F   8. T   9. T   10. F

Multiple Choice: 1. c   2. a   3. c   4. c   5. c   6. d   7. d   8. c   9. c   10. a   11. c

Reasoning and Expression:

1. One liter of 0.67 M $AgNO_3$
2. 0.0026 K .050%
3. (a) Increased temperature should increase solubility for a solid in a liquid (b) Increased temperature should decrease solubility of a gas in a liquid.
4. 500 g C_2H_5OH; 3700 g $C_{12}H_{22}O_{11}$
5. For the salt solution, measure out 58.5 g of salt and add to a container. Fill with water to exactly the 1 liter mark. For the sugar solution, measure out 342 grams of sugar and add to a container. Then measure out 1 kg of water. Add this to the container holding the sugar.
6. If the solute is not very water-soluble, then the solution will saturate quickly. It will also be dilute, since there will be very little solute in a relatively large amount of solvent. The solute would have a very low solubility; for example, 1 x 10^{-4} mol/L.

Chapter 15 Test

True - False: 1. T 2. T 3. T 4. T 5. T 6. T 7. T 8. T 9. F 10. F 11. T 12. F

Multiple Choice: 1. c 2. c 3. d 4. b 5. a 6. c 7. b 8. c 9. d 10. a 11. c 12. a

Reasoning and Expression:

1. (a) $HSO_4^- + H_2O \rightarrow SO_4^{2-} + H_3O^+$; conjugate acid - H_3O^+

(b) $CO_3^{2-} + NH_4^+ \rightarrow HCO_3^- + NH_3$; conjugate acid - HCO_3^-

(c) $HS^- + NO_2^- \rightarrow HNO_2 + S^{2-}$; conjugate acid - HNO_2

(d) $OH^- + H_2O \rightarrow H_2O + OH^-$; conjugate acid - H_2O
2. (a) 0.02 M (b) 1.7 (c) 4 g NaOH
3. NH_3 can accept a proton, as Bronsted describes a base:

$$NH_3 + H^+ \rightarrow NH_4^+$$

but it cannot donate an OH^- as Arrhenius' definition requires.
4. 10 g $CaCO_3$
5. $[Al^{3+}] = 0.107$ M, $[Cr_2O_7^{2-}] = 0.160$ M $Al_2(Cr_2O_7)_3 \rightarrow 2 Al^{3+} + 3 Cr_2O_7^{2-}$
6. (a) no (b) yes

(c) 0.5 ml of 1M NaOH in excess. 0.5 ml x $\frac{1 L}{1000 mL}$ x $\frac{1 mol}{L}$ = 5 x 10^{-4} mol NaOH

$[OH^-] = \frac{5.0 \times 10^{-4} mol}{0.0995 L}$ = 5 x 10^{-3} M

pOH = 2.3 → pH = 11.7

(d) 0.5 mL of 1M HI

Chapter 16 Test

True - False: 1. T 2. T 3. F 4. F 5. T 6. T 7. F 8. F 9. F 10. F 11. F

Multiple Choice: 1. b 2. c 3. d 4. a 5. a 6. a 7. b 8. a 9. d 10. a 11. d

Reasoning and Expression:

1. The temperature could be increased; the amount of N_2 or H_2 could be increased; the pressure could be increased; and the ammonia could be removed as it is being produced.

2. 0.95

3. 0.42%, 4.2% ionized

4. $[Ag^+] = 1.0 \times 10^{-16}$ M, $[S^{2-}] = 5.1 \times 10^{-17}$ M

5. No—with sugar on the bottom of the glass, the solution is already saturated. Adding additional sugar will simply add to the solid at the bottom. No greater concentration of dissolved particles will result.

6. (a) $HA \rightleftarrows H^+ + A^-$ $K_a = 1.8 \times 10^{-5} = \frac{[H^+][A^-]}{[HA]} = \frac{(x)(x)}{(1.0-x)} \simeq \frac{x^2}{1.0}$

$x = [H^+] = 4.2 \times 10^{-3} \rightarrow$ % ioniz. $= \frac{4.2 \times 10^{-3}}{1.0} \times 100 = 0.42\%$

(b) 1.3% (c) 4.2%

The % ionization increases. As the solution becomes more dilute, a greater proportion of the solute molecules will ionize under the influence of the water molecules that are present. They act as a Bronsted-Lowry base.

Chapter 17 Test

True - False: 1. F   2. F   3. F   4. T   5. T   6. T   7. T   8. F   9. T   10. T   11. F

Multiple Choice: 1. c   2. a   3. b   4. a   5. a   6. d   7. a   8. a   9. b   10. d   11. b   12. c   13. c

Reasoning and Expression:

1. (a) $3P + 5HNO_3 \rightarrow 3 HPO_3 + 5NO + H_2O$

(b) $S^{2-} + 8 OH^- + 4 Cl_2 \rightarrow SO_4^{2-} + 8 Cl^- + 4 H_2O$

(c) $2 MnO_4^- + 10 Cl^- + 16 H^+ \rightarrow 2 Mn^{2+} + 5 Cl_2 + 8 H_2O$

2. (a) Toward Ag (b) Zn (c) Toward the silver strip (d) $Zn \rightarrow Zn + 2e^-$; $Ag^+ + e^- \rightarrow Ag$

3. oxidizing agent: I_2, reducing agent: SO_3^{2-}

4. $P_4(0) < H_3PO_3(+3) < PF_5$ and P_4O_{10} and $Ca_3(PO_4)_2$(all +5)

5. $Al + Cu^{2+} \rightarrow Al^{3+} + Cu$; aluminum metal is being oxidized by the copper (II) ions, they in turn are being reduced; electrons are being transferred from Al to Cu^{2+}.

6. anode reaction: $Zn + 2 OH^- \rightarrow Zn(OH)_2 + 2 e^-$

cathode reaction: $MnO_2 + H_2O + e^- \rightarrow MNO(OH) + OH^-$

overall reaction: $Zn + 2MnO_2 + 2 H_2O \rightarrow Zn(OH)_2 + 2 MnO(OH)_{(s)}$

Chapter 18 Test

True - False: 1. T   2. F   3. T   4. F   5. T   6. T   7. T   8. F   9. F   10. T   11. F

Multiple Choice: 1. b   2. d   3. d   4. c   5. c   6. b   7. d   8. b   9. d   10. a   11. d   12. c   13. c

Reasoning and Expression:

1. (a) α (b) ^7Li (c) ^{14}C (d) n

2. (a) Its nuclear charge will decrease by 2 and its mass will decrease by 4 amu

(b) Its mass will increase by 2 amu (c) Its nuclear charge will decrease by 3 and its mass will decrease by 5 amu

3. 2 hours

4. (a) 52 (b) $\frac{52}{38} = \frac{26}{19} = 1.4$ (c) $^{88}_{36}Kr$ (d) Strontium is in the same chemical family as calcium and, therefore, reacts similarly. As a result, strontium will be found in bone tissue as is calcium.

5. Charge, 41. Mass, 90.

6. (a) The alpha particle has four times the mass and two times the positive charge of the proton.
(b) The proton and the neutron are equal in mass; the proton has a positive charge of 1, and the neutron is not charged.
(c) The proton is much more massive than the electron; they have equal amounts of electrical charge, though the charge is opposite; positive for the proton and negative for the electron.
(d) The alpha particle is twice as massive and twice as positively charged as the deuteron.
7. Electrons are held onto the atom with much less attractive force than nucleons are held together in the nucleus.

Chapter 19 Test

True - False: 1. T 2. T 3. F 4. F 5. F 6. F 7. T 8. T 9. F 10. T 11. T

Multiple Choice: 1. b 2. a 3. c 4. c 5. b 6. d 7. c 8. d 9. c 10. b 11. c

Reasoning and Expression:

1. (a)
$$CH_3-\underset{\underset{Cl}{|}}{\overset{\overset{Cl}{|}}{C}}-\underset{\overset{|}{CH_3}}{\overset{\overset{CH_3}{|}}{CH}}$$
(b)
$$CH_3-CH-\underset{\overset{|}{CH_3}}{\overset{\overset{CH_3}{|}}{C}}-CH_2-CH_2-CH_3$$ with CH_3 on the first CH

(c)
$$Cl-CH_2-\underset{\underset{Cl}{|}}{\overset{\overset{Cl}{|}}{C}}-CH_3$$

2. (a)
$$H-\underset{\underset{Cl}{|}}{\overset{\overset{H}{|}}{C}}-\underset{\underset{Cl}{|}}{\overset{\overset{H}{|}}{C}}-H$$
(b)
$$H-\underset{\underset{OH}{|}}{\overset{\overset{H}{|}}{C}}-\underset{\underset{OH}{|}}{\overset{\overset{H}{|}}{C}}-H$$
(c)
$$H-\overset{\overset{O}{\|}}{C}-O-CH_3 \quad +H_2O$$

3. Carbon's tetravalent bonding capacity allows for much branching and diversity in molecular structure.
4. (a) Alkenes contain a double bond and alkynes contain a triple. (b) Aldehydes are simple hydrocarbons; alcohols are compounds which also contain an -OH. (c) Aldehydes and carboxylic acids both contain a carbon-oxygen double bond, but acids have the general formula RCOOH and aldehydes have the formula RCHO.
5. 1-hydroxypentane $CH_3-CH_2-CH_2-CH_2CH_2OH$

2-hydroxypentane
$$CH_3-CH_2-CH_2-\underset{\overset{|}{OH}}{CH}-CH_3$$

3-hydroxypentane
$$CH_3-CH_2-\underset{\overset{|}{OH}}{CH}-CH_2-CH_3$$

1-hydroxy-3-methylbutane
$$CH_3-\underset{\overset{|}{CH_3}}{CH}-CH_2-CH_2-OH$$

6. Carbons can have a maximum of four covalent bonds. The four valence electrons that they have in s and p orbitals preclude more bonds than that. Each carbon in the molecule has more than four bonds and, thus, the proposed is doubtful.

Chapter 20 Test

True - False: 1. T   2. T   3. F   4. F   5. F   6. T   7. T   8. T   9. T   10. T   11. F

Multiple Choice: 1. a   2. c   3. c   4. a   5. a   6. c   7. c   8. c   9. c   10. b   11. c   12. c

Reasoning and Expression:

1. (a) Breaks a disaccharide into two monosaccharides (b) converts double bonds in oil and fat molecules to single bonds by adding hydrogen (c) hydrolysis of a fat or an oil using NaOH to form a soap and glycerol.
2. (a) Fatty acids are long chain carboxylic acids, whereas fats are esters of fatty acids and glycerol. (b) complex sugars are comprised of simple sugars, and (c) saturated fats have only carbon-carbon single bonds whereas unsaturated fats have some carbon double bonds.
3. Photosynthesis produces oxygen and sugar or starch from starting materials of water and carbon dioxide. Respiration does just the opposite. Photosynthesis occurs in plants; respiration occurs in animals.
4. 1.7×10^{22} molecules
5. Proteins are polymers comprised of amino acids; peptide is the name of the bond that holds the amino acids together to make the proteins, polypeptides are units of amino acids with up to about 40–50 amino acids. Polypeptides are comprised of shorter amino acid chains than proteins.

Answers to Test Questions (pages 77–268)

Chapter 1. Introduction

True - False:

1. T	2. T	3. F	4. T	5. T	6. F	7. T	8. T	9. T	10. F	11. T
12. T	13. T	14. F	15. T	16. F	17. F	18. T	19. F	20. T	21. T	22. T
23. F	24. T	25. T	26. F	27. T	28. F	29. F				

Multiple Choice: 1. d 2. b 3. c 4. c 5. c

Chapter 2. Standards of Measurements

True - False:

1. F	2. F	3. T	4. F	5. F	6. T	7. F	8. T	9. T	10. T	11. F
12. F	13. T	14. T	15. T	16. F	17. F	18. F	19. F	20. T	21. T	22. T
23. F	24. F	25. T	26. F	27. F	28. F	29. T	30. F	31. F	32. T	33. F
34. T	35. F	36. T	37. T	38. T	39. T	40. F	41. F	42. T	43. T	44. F
45. F	46. F	47. F	48. T	49. F	50. F	51. F	52. T	53. F	54. F	55. T

Multiple Choice:

1. c	2. a	3. d	4. c	5. d	6. b	7. b	8. a	9. c	10. b	11. c
12. d	13. a	14. c	15. b	16. a	17. d	18. b	19. c	20. b	21. d	22. b
23. c	24. d	25. d	26. c	27. c	28. c	29. c	30. b	31. b	32. a	33. c
34. a	35. a	36. d	37. c	38. b	39. d	40. b	41. a	42. c	43. b	44. a
45. c	46. d	47. d	48. b	49. c	50. a	51. a	52. a	53. b	54. a	55. a
56. c	57. b	58. d								

Matching: 1. (i) 2. (e) 3. (b) 4. (d) 5. (f) 6. (g) 7. (j) 8. (a) 9. (h) 10. (c)

Chapter 3. Classification of Matter

True - False:

1. T	2. F	3. F	4. T	5. F	6. F	7. F	8. T	9. F	10. T	11. T
12. F	13. F	14. T	15. T	16. T	17. T	18. F	19. T	20. F	21. T	22. T
24. T	25. T	26. T	27. T	28. F	29. T	30. F	31. T	32. F	33. F	34. F
35. F	36. T	37. F	38. T	39. T	40. F	41. T	42. T	43. F	44. T	45. T
46. F	47. F	48. T	49. T	50. T	51. F	52. F	53. F	54. T	55. F	56. T
57. T	58. F	59. F	60. F	61. F	62. T	63. F	64. F	65. T	66. T	67. F
68. T										

Multiple Choice:

1. b	2. a	3. b	4. a	5. b	6. d	7. b	8. c	9. a	10. d	11. d
12. b	13. a	14. b	15. d	16. a	17. b	18. d	19. c	20. a	21. c	22. a
23. b	24. c	25. b	26. c	27. d	28. c	29. a	30. d	31. c	32. a	33. a
34. c	35. a	36. b	37. b	38. d	39. d	40. b				

Matching: 1. (d) 2. (e) 3. (c) 4. (h) 5. (b) 6. (f) 7. (a) 8. (g) 9. (j) 10. (i) 11. (m)
12. (o)

Chapter 4. Properties of Matter

True - False:

1. T	2. T	3. F	4. F	5. F	6. F	7. T	8. T	9. T	10. F	11. T
12. F	13. T	14. T	15. T	16. T	17. F	18. T	19. T	20. F	21. F	22. T
23. T	24. F	25. F	26. F	27. F	28. T	29. T	30. T	31. T	32. T	33. F
34. T	35. T	36. F	37. T	38. T	39. F	40. T	41. F	42. F	43. T	44. T
45. T	46. F	47. F	48. F							

Multiple Choice: 1. c 2. c 3. d 4. c 5. d 6. d 7. a 8. d 9. c 10. a 11. a
12. a 13. a 14. c 15. b 16. b 17. d 18. d 19. b 20. a 21. d 22. a
23. a 24. d 25. a 26. b 27. a 28. a 29. d 30. d 31. c 32. b 33. c
34. a 35. d

Chapter 5. Early Atomic Theory and Structure

True - False: 1. F 2. T 3. F 4. T 5. T 6. F 7. T 8. T 9. F 10. T 11. T
12. F 13. F 14. F 15. T 16. F 17. F 18. T 19. F 20. F 21. T 22. T
23. F 24. F 25. T 26. T 27. T 28. F 29. T 30. T 31. T 32. T 33. F
34. F 35. F 36. F 37. T 38. T 39. F 40. T 41. F 42. T 43. T 44. T
45. T 46. F 47. F 48. F 49. T 50. F

Multiple Choice: 1. b 2. d 3. b 4. d 5. c 6. c 7. d 8. d 9. d 10. b 11. a
12. d 13. c 14. a 15. c 16. c 17. a 18. b 19. a 20. d 21. b 22. b
23. b 24. d 25. c 26. a 27. d 28. a 29. d 30. b 31. d 32. c 33. b
34. c 35. d 36. c

Matching: 1. (b) 2. (g) 3. (d) 4. (j) 5. (a) 6. (i) 7. (f) 8. (n) 9. (o) 10. (m)

Chapter 6. Nomenclature of Inorganic Compounds

True - False: 1. F 2. T 3. F 4. T 5. T 6. F 7. F 8. T 9. T 10. F 11. T
12. T 13. T 14. T 15. F 16. F 17. T 18. T 19. F 20. T 21. F 22. F
23. F 24. F 25. F 26. F 27. F 28. T 29. F 30. T 31. T 32. T 33. T
34. F 35. T 36. F 37. T 38. T 39. F 40. F 41. T 42. T 43. T 44. T
45. F 46. T 47. T 48. F 49. T 50. F 51. F 52. T 53. F 54. F 55. T
56. T 57. T 58. F 59. T 60. T 61. T 62. F 63. T 64. F 65. F 66. T
67. T 68. F 69. F 70. T

Multiple Choice: 1. c 2. d 3. b 4. a 5. d 6. a 7. c 8. d 9. c 10. a 11. b
12. b 13. c 14. c 15. a 16. b 17. c 18. a 19. b 20. c

Matching I: 1. (l) 2. (h) 3. (o) 4. (e) 5. (b) 6. (n) 7. (p) 8. (f) 9. (d) 10. (r)

Matching II. 1 . (a) 2. (h) 3. (f) 4. (q) 5. (m) 6. (o) 7. (e) 8. (i) 9. (k) 10. (j)

Matching III: 1. (b) 2. (m) 3. (c) 4. (e) 5. (n) 6. (g) 7. (h) 8. (j) 9. (1) 10. (d)

Completion:
1. H_2CO_3, carbonic acid
2. $Fe(NO_3)_2$, iron(II) nitrate
3. $AlCl_3$, aluminum chloride
4. $Ca_3(PO_4)_2$, calcium phosphate
5. $(NH_4)_2SO_4$, ammonium sulphate
6. Hg_2SO_3, mercury(I) sulfite
7. $Mg(ClO)_2$, magnesium hypohlorite
8. $KClO_3$, potassium chlorate
9. $NaClO_4$, sodium Perchlorate
10. Cu_2O, copper(I) oxide
11. $Ni(C_2H_3O_2)$, nickel(II) acetate
12. $Sn(CrO_4)_2$, tin(IV) chromate
13. $BaCr_2O_7$, barium dichromate
14. $KMnO_4$, potassium permanganate
15. Na_2S, sodium sulfide
16. LiI, lithium iodide
17. $Fe(HCO_3)_3$, iron(III) bicarbonate
18. $Cd(CN)_2$, cadmium cyanide
19. $Zn(OH)_2$, zinc hydroxide
20. $AgBr$, silver bromide

Chapter 7. Quantitative Composition of Compounds

True - False:

1. F	2. T	3. F	4. T	5. F	6. F	7. F	8. F	9. F	10. T	11. T
12. T	13. F	14. F	15. F	16. F	17. F	18. T	19. T	20. T	21. T	22. T
23. T	24. T	25. F	26. T	27. T	28. T	29. T	30. T	31. F	32. T	33. F
34. T	35. F	36. T	37. F	38. F	39. F	40. T	41. T	42. T	43. F	44. F
45. T	46. F	47. F	48. T	49. T (It cannot be simplified.)		50. T	51. T	52. F		

Multiple Choice:

1. d	2. a	3. b	4. d	5. c	6. c	7. b	8. d	9. c	10. d	11. a
12. a	13. d	14. a	15. c	16. d	17. c	18. c	19. b	20. a	21. a	22. b
23. b	24. c	25. a	26. c	27. c	28. a	29. a	30. d	31. c	32. c	33. b
34. d	35. d	36. d	37. c	38. a	39. a	40. c	41. a	42. d	43. c	44. a
45. b	46. d	47. a	48. a	49. b	50. a	51. d	52. d	53. c	54. b	55. a
56. b	57. b	58. c	59. b	60. c	61. b	62. b				

Chapter 8. Chemical Equations

True - False:

1. T	2. T	3. T	4. F	5. T	6. T	7. T	8. F	9. F	10. F	11. T
12. F	13. F	14. T	15. T	16. T	17. F	18. F	19. T	20. F	21. T	22. F
23. T	24. T	25. T	26. T	27. T	28. T	29. F	30. F	31. F	32. T	33. T
34. F	35. T	36. F	37. F	38. F	39. T	40. T	41. F	42. F	43. F	44. F

Multiple Choice:

1. c	2. b	3. d	4. a	5. a	6. b	7. b	8. c	9. d	10. a	11. b
12. c	13. d	14. a	15. b	16. d	17. c	18. a	19. a	20. c	21. a	22. b
23. d	24. b	25. d	26. d	27. a	28. b	29. c	30. a	31. b	32. b	33. b
34. b	35. c	36. b	37. c	38. b	39. d	40. d	41. d	42. c		

Completion: 1. $\rightarrow$ 2. $\rightleftarrows$ 3. $\uparrow$, *(g)* 4. $\downarrow$, *(s)* 5. *(l)* 6. Δ 7. + 8. *(aq)*

Chapter 9. Calculations from Chemical Equations

True - False:

1. T	2. T	3. F	4. T	5. F	6. T	7. T	8. F	9. T	10. T	11. F
12. T	13. F	14. F	15. F	16. F	17. F	18. T	19. F	20. F	21. T	22. T
23. F	24. T	25. F	26. F	27. F						

Multiple Choice:

1. a	2. a	3. c	4. d	5. d	6. c	7. c	8. a	9. a	10. d	11. c
12. d	13. b	14. a	15. c	16. c	17. a	18. a	19. c	20. c	21. b	22. c
23. c	24. a	25. c	26. d	27. b	28. d	29. d	30. d	31. d	32. a	33. a
34. b	35. b	36. c	37. d	38. c	39. d	40. b	41. b	42. a	43. c	44. c
45. b	46. c									

Chapter 10. Modern Atomic Theory

True - False:

1. T	2. T	3. T	4. F	5. T	6. F	7. T	8. F	9. T	10. F	11. T
12. F	13. T	14. T	15. F	16. F	17. T	18. T	19. T	20. F	21. F	22. T
23. F	24. T	25. F	26. F	27. F	28. T	29. T	30. T	31. F	32. T	33. F
34. F	35. T	36. T	37. F	38. T	39. F	40. F	41. F	42. T	43. F	44. F
45. T	46. F									

Multiple Choice: 1. c 2. c 3. c 4. a 5. c 6. a 7. a 8. b 9. a 10. d 11. b
12. d 13. c 14. d 15. b 16. c 17. c 18. d 19. b 20. d 21. d 22. b
23. a 24. c 25. c 26. d 27. c 28. d 29. b 30. b 31. b 32. d 33. d
34. b 35. d

Chapter 11. The Formation of Compounds from Atoms

True - False: 1. T 2. F 3. F 4. T 5. T 6. F 7. T 8. F 9. T 10. F 11. T
12. T 13. T 14. T 15. F 16. T 17. T 18. F 19. F 20. T 21. T 22. T
23. F 24. T 25. F 26. T 27. T 28. T 29. T 30. T 31. T 32. F 33. F
34. T 35. T 36. T 37. T 38. F 39. F 40. T 41. F 42. F 43. F 44. F
45. T 46. T 47. F 48. T 49. T 50. F 51. F 52. F 53. F 54. T 55. T
56. T 57. T 58. T 59. T 60. T 61. T 62. T 63. T 64. F 65. T 66. F
67. F

Multiple Choice: 1. a 2. b 3. c 4. b 5. d 6. c 7. b 8. b 9. b 10. b 11. a
12. d 13. a 14. d 15. a 16. d 17. b 18. d 19. b 20. a 21. b 22. c
23. a 24. c 25. a 26. d 27. b 28. c 29. c 30. d 31. c 32. c 33. d
34. d 35. b 36. b 37. c 38. c 39. d 40. d 41. b 42. d 43. b 44. a
45. d 46. c 47. c 48. c 49. b 50. d 51. b 52. a 53. c 54. c 55. a
56. d 57. d 58. c 59. c 60. a 61. d 62. a 63. d

Matching: 1. (h) 2. (d) 3. (d) 4. (i) 5. (e) 6. (l) 7. (f) 8. (g)

Chapter 12. The Gaseous State of Matter

True - False: 1. T 2. T 3. F 4. F 5. T 6. T 7. T 8. T 9. F 10. F 11. F
12. F 13. T 14. T 15. T 16. T 17. T 18. T 19. T 20. T 21. F 22. T
23. F 24. F 25. F 26. F 27. F 28. F 29. T 30. T 31. T 32. T 33. T
34. F 35. T 36. T 37. T 38. T 39. T 40. T 41. T 42. T 43. F 44. F
45. T 46. T 47. T 48. F 49. T 50. T 51. T 52. F 53. F 54. T 55. T
56. T 57. F 58. F 59. T 60. T 61. F 62. T 63. T 64. F 65. T 66. T
67. F 68. T 69. F 70. T 71. F 72. T 73. F 74. T 75. T 76. F 77. T
78. F 79. F 80. F 81. T

Multiple Choice: 1. c 2. d 3. a 4. c 5. c 6. c 7. d 8. b 9. b 10. a 11. d
12. d 13. d 14. d 15. b 16. a 17. a 18. a 19. c 20. c 21. c 22. c
23. b 24. a 25. d 26. c 27. b 28. d 29. b 30. a 31. d 32. b 33. c
34. b 35. a 36. b 37. b 38. a 39. b 40. c 41. d 42. c 43. b 44. d
45. a 46. c 47. d 48. b 49. c 50. b 51. a 52. b 53. a 54. a 55. a
56. c 57. d 58. c 59. d 60. d 61. d 62. b 63. c 64. d 65. a 66. a
67. d 68. d 69. b 70. b 71. b 72. a 73. b 74. d 75. d

Chapter 13. Water and the Properties of Liquids

True - False: 1. T 2. T 3. T 4. F 5. F 6. T 7. F 8. F 9. T 10. F 11. T
12. F 13. F 14. F 15. T 16. F 17. T 18. T 19. T 20. F 21. T 22. F
23. T 24. T 25. T 26. T 27. T 28. F 29. T 30. F 31. F 32. F 33. F
34. T 35. T 36. T 37. T 38. T 39. T 40. F 41. F 42. F 43. T 44. T
45. F 46. T 47. F 48. T 49. F 50. F 51. T 52. T 53. F 54. T 55. F
56. F 57. T 58. F 59. T 60. T 61. F 62. F 63. T 64. F 65. F 66. T

67. T	68. F	69. T	70. T	71. T	72. T	73. F	74. T	75. T	76. T	77. T
78. T	79. F	80. F	81. F	82. T	83. T	84. T	85. F	86. F	87. T	88. F
89. F	90. F	91. T								

Multiple Choice:

1. b	2. d	3. c	4. b	5. a	6. b	7. c	8. a	9. a	10. b	11. d
12. d	13. b	14. d	15. c	16. b	17. c	18. a	19. a	20. d	21. b	22. c
23. d	24. c	25. d	26. c	27. d	28. d	29. b	30. c	31. c	32. c	33. a
34. b	35. b	36. c	37. c	38. d	39. c	40. c	41. b	42. c	43. a	44. c
45. b	46. a	47. c	48. a	49. c	50. d	51. b	52. b	53. b	54. a	55. a
56. a	57. c	58. c	59. c	60. d	61. b	62. c	63. d	64. b	65. a	66. d

Matching:

1. (e)	2. (k)	3. (d)	4. (f)	5. (j)	6. (b)	7. (i)	8. (g)	9.(h)

Chapter 14. Solutions

True - False:

1. T	2. F	3. T	4. F	5. F	6. T	7. F	8. T	9. T	10. T	11. T
12. F	13. T	14. T	15. F	16. T	17. F	18. T	19. F	20. F	21. T	22. T
23. F	24. T	25. T	26. F	27. T	28. T	29. F	30. T	31. F	32. T	33. F
34. T	35. T	36. T	37. T	38. F	39. F	40. T	41. T	42. T	43. T	44. F
45. T	46. F	47. T	48. F	49. T	50. T	51. T	52. T	53. F	54. T	55. F
56. T	57. T	58. F	59. T	60. F	61. F	62. F	63. F	64. T	65. T	66. F
67. T	68. T	69. F	70. F	71. T	72. F	73. F	74. F	75. F	76. T	77. F

Multiple Choice:

1. a	2. c	3. b	4. b	5. a	6. d	7. d	8. c	9. c	10. b	11. a
12. d	13. a	14. a	15. c	16. a	17. c	18. d	19. a	20. a	21. d	22. b
23. c	24. b	25. b	26. b	27. c	28. a	29. a	30. c	31. a	32. c	33. c
34. b	35. d	36. d	37. a	38. c	39. c	40. d	41. a	42. a	43. b	44. a
45. b	46. d	47. c	48. d	49. a	50. b	51. a	52. c	53. c	54. b	55. d
56. d	57. b	58. c	59. c	60. a	61. b	62. a	63. b	64. b	65. d	66. d
67. b	68. d	69. b	70. c	71. d						

Matching:

1. (m)	2. (a)	3. (i)	4. (e)	5. (k)	6. (g)	7. (c)	8. (j)	9. (h)	10. (f)

Chapter 15. Acids, Bases, and Salts

True - False:

1. T	2. F	3. T	4. F	5. T	6. T	7. T	8. F	9. F	10. T	11. F
12. T	13. F	14. F	15. F	16. T	17. T	18. T	19. T	20. T	21. T	22. T
23. T	24. F	25. T	26. F	27. T	28. T	29. T	30. T	31. F	32. F	33. T
34. T	35. F	36. F	37. T	38. F	39. T	40. F	41. F	42. F	43. F	44. F
45. T	46. F	47. F	48. T	49. T	50. T	51. T	52. F	53. T	54. F	55. T
56. T	57. F	58. F	59. T	60. F	61. T	62. T	63. T	64. F	65. F	66. F
67. T	68. T	69. T	70. F	71. T	72. T	73. F	74. F	75. T	76. F	77. F
78. F	79. T	80. F	81. T	82. T	83. F	84. F	85. T	86. T	87. T	88. F
89. T	90. F	91. T	92. T	93. T	94. F	95. F	96. F	97. F	98. T	99. F
100. T	101. F	102. F	103. F	104. T						

Multiple Choice:

1. c	2. a	3. c	4. c	5. a	6. a	7. c	8. c	9. a	10. c	11. b
12. b	13. a	14. b	15. b	16. d	17. a	18. c	19. c	20. a	21. c	22. b
23. c	24. c	25. d	26. d	27. b	28. b	29. c	30. a	31. b	32. c	33. d
34. b	35. a	36. b	37. a	38. d	39. b	40. c	41. c	42. b	43. c	44. b
45. d	46. d	47. a	48. b	49. c	50. d	51. b	52. c	53. d	54. a	55. c

56. d   57. c   58. d   59. a   60. d   61. c   62. c   63. b   64. c   65. b   66. d

Matching: 1. (i)   2. (g)   3. (n)   4. (k)   5. (f)   6. (e)   7. (m)   8. (c)   9. (j)   10. (h)

Chapter 16. Chemical Equilibrium

True - False:

1. T	2. T	3. F	4. T	5. T	6. T	7. F	8. T	9. T	10. T	11. T
12. F	13. T	14. F	15. T	16. T	17. F	18. F	19. T	20. F	21. F	22. T
23. T	24. F	25. T	26. F	27. T	28. T	29. F	30. F	31. F	32. T	33. F
34. T	35. F	36. F	37. F	38. F	39. T	40. T	41. T	42. F	43. F	44. T
45. F	46. T	47. F	48. F	49. F	50. F	51. T	52. T	53. T	54. F	55. T
56. F	57. T	58. T	59. T	60. F	61. F	62. F	63. T	64. F	65. T	66. T
67. T	68. T	69. F	70. T	71. T	72. F	73. T	74. T	75. F	76. F	77. T
78. T	79. F	80. T	81. F	82. T	83. T	84. F	85. F	86. T	87. T	88. F
89. F										

Multiple Choice:

1. d	2. d	3. b	4. d	5. a	6. b	7. c	8. b	9. a	10. a	11. c
12. c	13. a	14. a	15. b	16. c	17. d	18. c	19. b	20. d	21. b	22. b
23. b	24. c	25. d	26. b	27. c	28. c	29. c	30. b	31. a	32. d	33. a
34. d	35. c	36. c	37. c	38. b	39. b	40. b	41. a	42. c	43. c	44. d
45. a	46. d	47. c	48. b	49. b	50. a	51. d	52. c	53. b	54. c	55. d
56. a	57. c	58. c	59. a	60. b	61. b	62. c	63. c	64. b	65. b	66. a
67. b	68. c	69. d	70. b	71. d						

Chapter 17. Oxidation-Reduction

True - False:

1. T	2. F	3. T	4. T	5. F	6. F	7. T	8. F	9. T	10. T	11. F
12. F	13. T	14. T	15. F	16. T	17. T	18. F	19. F	20. F	21. T	22. F
23. T	24. F	25. T	26. F	27. T	28. T	29. T	30. F	31. T	32. T	33. F
34. T	35. T	36. T	37. T	38. F	39. T	40. F	41. T	42. F	43. F	44. T
45. T	46. T	47. T	48. F	49. T	50. T	51. T	52. F	53. T	54. F	55. F
56. F	57. F	58. T	59. F	60. F	61. T					

Multiple Choice:

1. c	2. a	3. b	4. d	5. d	6. d	7. c	8. b	9. b	10. b	11. b
12. b	13. c	14. b	15. b	16. c	17. d	18. a	19. c	20. b	21. a	22. b
23. d	24. b	25. b	26. d	27. a	28. b	29. d	30. c	31. b	32. c	33. a
34. b	35. b	36. a	37. c	38. d	39. b	40. a	41. a	42. d	43. a	44. c
45. c	46. d	47. b								

Balancing Equations:

1. $2\ MnSO_4 + 5\ PbO_2 + 3\ H_2SO_4 \rightarrow 2\ HMnO_4 + 5\ PbSO_4 + 2\ H_2O$

2. $Cr_2O_7^{2-} + 14\ H^+ + 6\ Cl^- \rightarrow 2\ Cr^{3+} + 7\ H_2O + 3\ Cl_2$

3. $2\ MnO_4^- + 5\ AsO_3^{3-} + 6\ H^+ \rightarrow 2\ Mn^{2+} + 5\ AsO_4^{3-} + 3\ H_2O$

4. $4\ Zn + NO_3^- + 6\ H_2O + 7\ OH^- \rightarrow 4\ Zn(OH)_4^{2-} + NH_3$

5. $2\ KOH + Cl_2 \rightarrow KCl + KClO + H_2O$

6. $4\ As + 3\ ClO_3^- + 6\ H_2O + 3\ H^+ \rightarrow 4\ H_3AsO_3 + 3\ HClO$

7. $Cl_2O_7 + 4\ H_2O_2 + 2\ OH^- \rightarrow 2\ ClO_2^- + 4\ O_2 + 5\ H_2O$

8. $6\ H^+ + 2\ HNO_3 + 6\ S_2O_3^{2-} \rightarrow 2\ NO + 4\ H_2O + 3\ S_4O_6^{2-}$

9. $16\,H^+ + 2\,MnO_4^- + 5\,C_2O_4^{2-} \rightarrow 2\,Mn^{2+} + 10\,CO_2 + 8\,H_2O$

10. $2\,HNO_2 + H_2S \rightarrow 2\,NO + 2\,H_2O + S$

Chapter 18. Nuclear Chemistry

True - False:	1. F	2. F	3. F	4. T	5. T	6. F	7. T	8. T	9. F	10. F	11. T
	12. T	13. T	14. F	15. T	16. F	17. T	18. T	19. F	20. F	21. T	22. F
	23. T	24. T	25. T	26. F	27. T	28. T	29. T	30. T	31. T	32. T	33. T
	34. T	35. F	36. T	37. T	38. T	39. T	40. T	41. T	42. T	43. T	44. T
	45. T	46. T	47. T	48. T	49. T	50. F	51. F	52. T	53. F	54. T	55. T
	56. T	57. T	58. T	59. T	60. F	61. T	62. T	63. T	64. T	65. T	66. T
	67. T	68. T	69. T	70. T	71. T	72. F	73. T	74. F	75. F	76. T	77. F
	78. F	79. T	80. F								

Multiple Choice:	1. a	2. c	3. a	4. b	5. b	6. a	7. d	8. c	9. d	10. c	11. c
	12. a	13. b	14. b	15. c	16. a	17. a	18. a	19. d	20. c	21. b	22. d
	23. d	24. d	25. d	26. c	27. d	28. c	29. b	30. d	31. d	32. a	33. b
	34. a	35. a	36. c	37. c	38. d	39. b	40. b	41. a	42. b	43. a	44. d
	45. d	46. b	47. d	48. b	49. a	50. d	51. a				

Matching:	1. (d)	2. (a)	3. (e)	4. (h)	5. (b)	6. (g)	7. (c)	8. (f)

Chapter 19. Introduction to Organic Chemistry

True - False:	1. T	2. T	3. F	4. F	5. T	6. F	7. T	8. F	9. F	10. T	11. F
	12. T	13. F	14. T	15. F	16. T	17. T	18. T	19. F	20. T	21. T	22. T
	23. T	24. T	25. F	26. T	27. T	28. F	29. F	30. T	31. T	32. F	33. T
	34. F	35. F	36. T	37. F	38. F	39. T	40. T	41. T	42. T	43. T	44. T
	45. T	46. T	47. T	48. T	49. F	50. T	51. F	52. F	53. T	54. T	55. T
	56. F	57. T	58. T	59. T	60. T	61. F	62. F	63. F	64. F	65. T	66. F
	67. F	68. T	69. T	70. T	71. F	72. T	73. T	74. F	75. F	76. F	77. F
	78. T	79. T	80. T	81. F	82. T	83. T	84. F	85. F	86. F	87. T	88. T
	89. T	90. T	91. T	92. T	93. T	94. T					

Multiple Choice:	1. b	2. a	3. c	4. b	5. a	6. c	7. b	8. b	9. b	10. c	11. a
	12. d	13. b	14. d	15. c	16. c	17. b	18. c	19. b	20. b	21. a	22. c
	23. b	24. d	25. c	26. b	27. a	28. a	29. c	30. b	31. b	32. a	33. c
	34. c	35. c	36. d	37. a	38. d	39. d	40. b	41. c	42. b	43. c	44. b
	45. d	46. c	47. c	48. d	49. b	50. c	51. d	52. c	53. a	54. a	55. c
	56. b	57. d	58. c	59. b	60. d	61. b	62. c	63. c	64. c	65. d	66. c
	67. b	68. b	69. b	70. d	71. d	72. b	73. c	74. c	75. c	76. d	77. b

Chapter 20. Introduction to Biochemistry

True - False:	1. F	2. F	3. T	4. T	5. F	6. T	7. T	8. T	9. T	10. F	11. T
	12. T	13. T	14. F	15. T	16. T	17. F	18. T	19. T	20. F	21. F	22. T
	23. F	24. T	25. T	26. T	27. F	28. T	29. T	30. T	31. F	32. F	33. T
	34. T	35. F	36. F	37. T	38. T	39. T	40. T	41. T	42. T	43. T	44. F
	45. T	46. T	47. F	48. T	49. T	50. T	51. T	52. F	53. F	54. T	55. F
	56. T	57. T	58. T	59. T	60. T	61. T	62. T	63. T	64. F	65. T	66. T
	67. F	68. T	69. F	70. T	71. T	72. T	73. F	74. T	75. F	76. F	

Multiple Choice:

1. b	2. b	3. a	4. b	5. b	6. d	7. a	8. a	9. c	10. a	11. d
12. b	13. a	14. d	15. a	16. c	17. b	18. b	19. c	20. a	21. d	22. c
23. c	24. c	25. d	26. b	27. b	28. a	29. d	30. a	31. c	32. d	33. a
34. c	35. d	36. d	37. b	38. b	39. a	40. a	41. b	42. d	43. b	44. d
45. d	46. d	47. c	48. d	49. a	50. c	51. d				

Multiple Choice:

1. b	2. b	3. a	4. b	5. b	6. d	7. a	8. a	9. c	10. a	11. d
12. b	13. a	14. d	15. a	16. c	17. b	18. b	19. c	20. a	21. d	22. c

Printed in the United States
36987LVS00001B/185